THE DISUNITY OF AMERICAN CULTURE

THE DISUNITY OF AMERICAN CULTURE

SCIENCE, RELIGION, TECHNOLOGY AND THE SECULAR STATE

JOHN C. CAIAZZA

Routledge
Taylor & Francis Group
LONDON AND NEW YORK

First published 2013 by Transaction Publishers

2 Park Square, Milton Park, Abingdon, Oxfordshire OX14 4RN
711 Third Avenue, New York, NY 10017

Routledge is an imprint of the Taylor & Francis Group, an informa business

First issued in paperback 2017

Library of Congress Catalog Number: 2012035093

Library of Congress Cataloging-in-Publication Data

Caiazza, John.
The disunity of American culture : science, religion, technology, and the secular state / John C. Caiazza.
p. cm.
ISBN 978-1-4128-5168-8
1. United States--Civilization--21st century. 2. Politics and culture--United States. 3. Religion and politics--United States. 4. Technology and state--United States. 5. Science--United States--Political aspects. 6. Secularism--United States. I. Title.
E169.Z83C35 2013
973.93--dc23

2012035093

ISBN 13: 978-1-4128-5168-8 (hbk)
ISBN 13: 978-1-138-51572-7 (pbk)

Dedicated to my sisters and brothers—Stephen†, Marianne,
Matt, Nick, Michael†, Rosemary, Joanne, and Donald

Contents

Acknowledgments

Parts of some of the chapters in this text have appeared previously in the following journals in a different form:

Intercollegiate Review, Chapters 5 and 8

Modern Age, Chapters 6, 11, and 14

Political Science Reviewer, Chapters 9 and 10

Zygon, Chapter 14

Homiletic and Pastoral Review, Chapter 12.

My thanks to their respective editors.

The author also wishes to thank three former editors who have been instrumental in the development and publication of the ideas presented in this text: George Panichas, longtime editor of *Modern Age,* and Irving Louis Horowitz, founding editor of Transaction Publishers, both now sadly deceased; also George W. Carey, former editor of *The Political Science Reviewer.* The author wishes to express his appreciation to the Managing Editor of Transaction Publishers, Jennifer Nippins, who has overseen to one degree or another, the transition of three texts from hard copy to MS Word files to book form published by Transaction. Finally, the author, as always, thanks his wife, Jo, and son, Paul, for help both essential and inexpressible.

"The individualism of modern society and the increasingly rapid and disruptive rate of social change brings about a situation in which for increasing numbers there is no over-all shape to the moral life but only a set of apparently arbitrary principles inherited from a variety of sources."

—Alasdair MacIntyre

"It is mere vanity to wish to distinguish a technique as good or bad according to its end. Whether technique acts to the advantage of a dictator or of a democracy, it makes use of the same weapons, acts on the individual and manipulates his subconscious in identical ways, and in the end leads to the formation of exactly the same type of human being . . . the well-kneaded citizen."

—Jacques Ellul

"The question of the man who philosophizes about history, that is, of the man who looks at the whole and the roots of things, runs: What is it that really takes place there?—Now, if someone were to answer: The disintegration of a culture takes place, or: The rise of a world empire occurs, or: Economic Development comes about, or: A class conflict is enacted—in such answers the philosophic question about what really is and, in the last analysis, happens in history would manifestly remain unappeased. What really and in the deepest analysis happens in history is salvation and disaster."

—Josef Pieper

Chapter 1

Dissonant Themes in American Culture

That there have been dissonances in American culture is not a new observation. This book is an account of contemporary American culture at a time when a different set of forces is influencing it than did in the past, altering it now to the point of near incomprehensibility. In our day, issues of identity—sexual and racial—have a serious effect on our culture and our politics, but even more influential is the hotly debated question of what degree of support the state is morally obligated to give the individual, financially and in other ways. Meanwhile, the popular culture is seemingly dominated by the industrialization of entertainment, going far beyond what radio and television previously offered, to where it threatens to overwhelm our personal selves by enticing us to identify with fictional creations of programmers and scriptwriters. Interaction with real people is replaced by interaction with computer images; meanwhile, identification with real people is increasingly based on emphasizing the differences between ourselves and other people, making it increasingly difficult to create overarching bonds among American citizens. Our social differences are now reaching a level of fractiousness that has resulted in the "polarization" of our politics and its consequent inability to solve the immense problems that the nation now faces.

Because of the complexity of American culture at the present time, it is unlikely that the whole can be encompassed in one overall account, so one particular approach must be selected. The general approach taken in this book begins by analyzing American culture in terms of a debate that first arose in the nineteenth century: the apparent conflict between religion and science; but, after treating several allied subjects, it ends with presentation of the issue of the expanding power of the new secular state. While at the end of this introductory chapter the reader will find a summary of the book, it might be more helpful to get a sense of it to understand that it weaves four themes together. If we think of the book as if it were a piece of music, however, we find that the themes do not make up a harmonious whole, but are in fact a dissonance, a disharmony that implies a widening conflict in the culture. In any case, there are four themes to be examined.

Science Replacing Religious Belief

As understood here, science has, in large part, replaced religion as the primary culture former in the contemporary world. "Religion," however, is a general word, referring more than usual among general terms to a very wide and possibly incoherent collection of phenomena. "Religion," when used in any discussion of secularism in the West, however, refers more particularly to the Christian religion; for it is the Christian religion, manifested in its authority, ubiquity, and power, that dominated its culture and politics from Europe's founding until the middle of the twentieth century in America. In its various faces and social forms, it is this Christian religion that is being replaced, in large part, as the culture former of Europe and the West generally, including America. The Christian church is the infamy that Voltaire and now Dawkins, Harris, Hitchens, and Dennett claim should be wiped from the earth. (It should be noted that a major part of the inspiration for this new atheism is not only opposition to Judeo-Christianity, but blatant fear of current Islamic expansionism. The violence characteristic of Islamist expansionism stains the reputation of all religion, which is thus understood in general as encouraging oppressive violence against nonbelievers without distinction.[1])

It is not merely the power of the Christian churches in their various social manifestations that are being relegated; it is Christianity in a philosophical sense as well, as a mode or form of thinking that is being systematically undermined, and here attention must be paid not only to Christianity but to its ideological forebear, Judaism. For it is not merely a sentimental ecumenism to speak of Judeo-Christianity in this context or because of guilt over Christian mistreatment of the Jewish people in past times. Rather, Judaism and Christianity have a shared attitude toward the relationship of ideal reality, represented by the figure of the Lord God, and of physical reality as lived day-to-day, for in Judeo-Christian terms, there is a definite, identifiable portrayal of the relationship that has implications and consequences historically and philosophically for modern science; and this connection, in turn, affects what happens as modern science becomes the agent of secularizing Western culture. That attitude is reflected in the fact that in both the Old and the New Testament, God or his agents (such as prophets, angels, or Jesus) are intimately concerned with the affairs of human beings and that the divine order is exactly understood as commanding, judging, guiding, inspiring, and loving the creatures He has made, ordering all reality to the good of the human race. The drama of the Judaic and Christian accounts of God's presence on the earth might be dismissed as merely one among a plethora of myths, not unlike those of the ancient Assyrians, Greeks, Incas, Chinese, and so on. Yet the Judeo-Christian account is different in the exact manner in which the realm of the divine interacts with that of the plane of earthly existence; as a result, its relegation in the name of modern science has important consequences for society, ethics, and science itself.

How Science Influences Culture

Asserting that science is a major cause, if not the exclusive one, of the contours of contemporary culture, what are its particular areas or forms of influence? That is, what are its entry points into Western culture? Basically there are two: the impact of scientific technology on human life and the cultural influence of scientific theories. Technology has manifestly improved our lives; usually, the examples given have to do with railroads and automobiles, or space technology, or electromagnetic communication, but the best example may be from medical research because it has extended the human life span almost double in the past century and a half (from roughly forty years to eighty) and made human life more comfortable and less fraught with fear. This realization in turn encourages a positive attitude toward the sciences in general, because it has brought about such medical and technological benefits. But this positive attitude is directed not to scientific theories such as the germ theory of disease or the theory of the electromagnetic field but rather induces a general belief that science in the form of scientific technology provides the most effective approach to solving the problems, salving the hurts, and providing the hope for the future of the human race.

The other form of scientific influence is in the cultural effect of the actual theories of science of which the outstanding contemporary examples are evolution and the big bang. Evolution is assumed not only to be able to explain the origin of life-forms (species) but also to give the underlying cause of all life, so that no other explanation of the major aspects of life—including *human* life—is necessary. Evolution is perceived by its more avid proponents to preclude the doctrines of religion, the explanations of philosophy, the insights of literature, and the traditions drawn from millennial human experience. Significant examples of influential scientific theories come from the physical as well as the biological sciences; while one could mention the influence of Newton's mechanical theory in the eighteenth century, the most notable contemporary example is the big bang account of the origin of the universe. The big bang theory has entered popular culture to the extent that it is the name of a popular television program whose theme song begins, "The universe was in a hot, dense state." The idea that by means of the big bang theory, science has explained the origin of the universe and thereby its ultimate meaning is becoming pervasive in the West (and possibly throughout the world); and it is feared by religious believers, for the big bang theory is assumed to replace the biblical account of the universe's origin, which, in Genesis, was accomplished by the word of God.

The approved story of the replacement of the Judeo-Christian religion by modern science is what can be called a "Whig history"—in this case, perhaps the essential Whig history: that of progress over false superstition, bigotry, and ignorance by the forces of untrammeled intellect, systematic observation, and pure intentions. This hardy, Victorian view of what Taylor calls a process of "subtraction" (i.e., of various dogmas proclaimed by religion being removed as scientific knowledge progresses) is largely false, however.[2] A basic assumption of the Whig view of science history is that the role of religion is essentially *explanatory*: that thunder and lightning, for example, need no longer be explained as the manifestation of the anger of Thor or Jupiter; that human disease need no longer be explained as the invasion of evil spirits into the body; or that the existence of the universe need no longer be explained by God's creative action. E. O. Wilson, the most prominent of contemporary evolutionists, provides a particularly hubristic expression of this view.[3]

A kind of temptation accompanies the progress of science. Astronomy serves as an example, because it is the first modern science having developed from a merely descriptive and observational study, of the positions of the stars and their paths over the year, to a modern science in which general, mathematical laws are used to explain and predict their behavior. This happened in the fifteenth and sixteenth centuries with the development of the Copernican description of the solar system and the Galilean use of mathematical laws to explain motion. The use of an abstract model prescinds from actual observation in the case of Copernican astronomy, in which the theoretical observer has to see himself looking down on the system of sun and attendant planets as they whirl about it. This means abstracting oneself from the surface of the Earth where we live and where in fact the sun *does* appear to revolve about us every twenty-four hours, rising in the east and setting in the west each day. Yet in abstracting the theoretical viewer, taking him up from the surface of the earth where he actually lives, up beyond the sky and into outer space where he looks down on the solar system, we have created an illusion and put ourselves as observers in the way of hubristic temptation. To look at astronomical charts and calculate, as Kepler and Newton did, is to lead less intelligent people (and the vast bulk of the human race is less intelligent than either figure) to infer that they not only understand the solar system but, in a certain manner, control it.[4]

It may seem as if that the temptation to imagine ourselves as godlike is a small price to pay for having attained a central truth about the place of human beings in relation to the astronomical universe. But while the price may seem negligible at the level of astronomical phenomena, as I will attempt to demonstrate in the second part of this book, it is less so at the level where *life itself* is the object of scientific investigation—and much less so when the causes and judgment of human behavior are. At the levels of biology and social science, the price of the advance of scientific understanding is much higher, because at these levels, the temptation to hubris leads to the denigration of the value of life itself, mechanistic accounts of human behavior that are caricatures, and atomistic accounts of social interaction that become the rationale for totalitarianism.

Finally, explaining the influence of science upon Western culture by means of both its technological results and its theories, however, implies a dissonance, for the reliance on technological applications for an understanding of science causes people to ignore the relevance of scientific theory. The result is that modern science is assumed to be something (what exactly is not specified) that produces goods that provide comfort and security for the middle class, rather than an intellectual process of discovery. The further consequence, as explained in Part IV, is that the "decline of physical theory" reinforces another influential cultural meme: postmodern critical theory, which is really a form of epistemological relativism. Together, this has led currently to the belief among educated elites that truth is no longer applicable to human concerns, since the assumed acquisition of truth produces hegemonic results. But epistemological relativism leads to moral relativism, which is highly problematic for the continued sustenance of democratic politics.

Secularization Theory

There is a large literature on secularization, which extends back to the nineteenth century. Ernst Troeltsch made the distinction between "Church civilization" and "Modern civilization," indicating a clear separation between the medieval and the post-medieval cultures based on the presence of the Church in day-to-day life and the subsequent lack thereof. Ferdinand Tonnies made the distinction between two types of social organization, "Gemeinschaft" and "Gesellschaft," that is the former characteristic of premodern society that was organic in which each separate aspect of society was organized into an ideological and social whole, and the modern setting that is characterized by individuals making contracts and having to make a presentation of self and take a stance regarding each distinctive aspect of society, between unity and disparateness, so to speak. Max Weber famously made the connection between the rise of Protestantism and capitalism and also characterized formally for the first time the nature of leadership styles and, ominously, the nature of bureaucracies that now dominate all areas of modern social life.

In a certain sense, these attempts to characterize the differences between medieval and modern life constitute the origins of modern scientific sociology, as if the felt life of modern civilization could be appreciated or evaluated only by means of contrast with what went before. There is present, in these early attempts, then, a kind of nostalgia for a time that, because it is outside the range of actual memory, has to be reconstructed—an artificial memory, as it were, for certainly the actual life of the Middle Ages—which was rent by war between factions; hostility among religious movements; and a common life characterized by disease, unremitting labor, early death, and illiteracy, at a time when the popes fielded armies against emperors and feudal lords, and when Aquinas, now seen as the quintessential representative of high medieval culture, had his "heretical" books burned by the bishop of Paris—was not a time of social or ideological unity. Yet it must *seem* so, if only because modern life is so constantly and insistently characterized by our inability to see life and see it whole, because there are so many competing social pressures that do not allow us even to rest for a moment and gather up our resources. And so we, as did the early sociologists, invent a time when life was easier because the pressures were not so intense—a time when life was simpler. This extends not just to a simplistic reconstruction of the Middle Ages but to earlier generations in our own era. Thus the cliché that things were "simpler" in our parents' generation, when all they had to worry about was Hitler or the Cold War or Civil Rights or the Great Depression. The irony, of course, is that things were not any simpler or easier for previous generations than they were in the Middle Ages.

Another aspect of this early sociology, a characteristic that continues to this day, is that it was supposedly based on the personal stance of the scholars as one of being distanced from what they were studying and writing about. That is, there was and is—characteristically of the social sciences in general as well as specifically of sociology—the claim of neutrality as if it is not a matter of personal concern to the scholar whether or not secularization is taking place, for the concern of the social scientist must be to distance himself from whatever emotions or convictions he or she may have on the issue in order to dispassionately expose all its elements, explain the process by which it takes place, and put it in a general context. This formal neutrality becomes a heavy cross to bear, however, when discussing such things as racism or neo-Nazism or hyperinflation or sex slavery, for in such cases where injustice and immorality are dominant, then a stance of neutrality is much more difficult to adopt and, to outside observers, will probably look as if the scholar lacks not only feelings but moral judgment as well. But even in subjects where injustice or immorality are not so apparent, the social scientist has been attracted to it because if nothing else, he or she is interested in it for discernible reasons, perhaps even feeling strongly about it. Anna Freud did not write about children because she had no feelings about them; Gunnar Myrdal did not write about African Americans because he was neutral on the subject of Jim Crow and segregation; Alfred Kinsey did not write about the sex habits of Americans because he had no personal interest in sexual behavior. Inevitably, the stance of formal neutrality of sociology and all of the social sciences can become a kind of pretense, as if with a wink and a nudge, we know that the scholars *do* have an opinion about what they are studying; and, further, that it has not only slipped unnoticed into their research but in fact has largely guided it.

With regards to secularization, it appears as if the nineteenth-century scholars were looking at it not only as a cultural process that they had newly discovered was taking place, but, in effect, as though they approved of it. Surely, their studies did not lead them or others to condemn secularization, for in the heavily Protestant culture in which they were writing, the identification of the Middle Ages with the Roman Catholic Church practically guaranteed that secularization would be seen as a good thing, as a freeing up of Western civilization from the judgmental priests, the feudal economy, the crabbed theology, the power of the papacy, and the stultifying class structure. Attacks on medieval culture came also from the notion that science is the key to cultural progress and that medieval theology is the enemy of scientific reason: the Galileo case was inevitably referred to in this argument. These characterizations also took place in the nineteenth century, when the appearance of the inevitable progress of Western civilization was unchallenged, except by a few poets and literary critics. The arrival of Darwinism put in place the idea that scientific reason was finally overcoming older notions, classical as well as medieval, of the superior place of the human race in God's created universe and of the nature of man as distinct from the rest of the animals due to our ability to think abstractly and make moral decisions.

It is also the case, however, that by the middle of the twentieth century, as Western civilization, particularly in Europe, reached a point of self-destruction, secularization was being looked at by cultural critics such as T. S. Eliot, Paul Tillich, and Henri de Lubac as a cause of Western cultural self-immolation.

Secularism in American Culture

Secularism is the most significant moral movement in America today, penetrating every organ of civil life, increasingly relegating religion to a restricted place in society and reflective of a worldwide development. As an idea and a policy, it overwhelms any attempt to forestall it or even, in effect, to *identify* it. The idea is so pervasive that it does not seem to its exponents necessary to defend it, for it seems as if it is only a matter of common sense or the latest development of inevitable historical progress. Secularism is, as it were, a tsunami, overtaking all walls and breaches erected to contain its force, even as it washes over ideals of transcendence and public examples of religious faith that have actuated American life from its beginning. Thus the mere mention of "God" in public, much less the expression of the Ten Commandments carved in stone outside of a courthouse, is deemed too explicit an imposition of religious belief on the public order. Traditional holidays such as Christmas and Thanksgiving must be evacuated of any mention of their religious origins—neutralized, as it were, to give the impression of a flat plain of public attention in which any projecting reference to a specific religious event or ideal is not merely *not mentioned* but unmentionable. The secularization of American (and European) society has accelerated to the point where it seems inevitable and desirable and beyond discussion.

This constitutes, it may be said, a "fight for *the soul of the West.*" Necessarily, studies and commentaries—historical, sociological, and personal—of which three may be mentioned here, now chase this phenomenon. First, there is Charles Taylor's *A Secular Age*, a truly massive book of nearly eight hundred pages of text, published in a well-wrought edition by Harvard University Press.[5] Given the high reputations of both Taylor and Harvard, it appears to be an attempt to be the magisterial, final word on the phenomenon of modern secularism. (However, this book has not had the influence that Taylor's previous book, *Sources of the Self*, has had among contemporary social scientists.) *A Secular Age* is, as it were, an overachieving spider's web, for any aspect of secularism—from past or present, from literature, philosophy, or science—whenever it is touched by Taylor, gives off vibrations that he then diligently tracks down and describes. He is an intellectual historian who cannot seem to put down his pen (or detach from his keyboard) at any point, but in the end, we can detect that Taylor is not in tune with the new secularism and warns against the rejection of religious doctrines and disciplines—what he calls the process of "excarnation," by which he appears to mean the dissolution of the intellectual bonds that connect the ideal and the realm of ideas from that of empirical study and material effects. Taylor's book does in fact constitute a remarkable history of secularism, but it is also diffuse, and no topic that Taylor explores is left without mention of other intellectual and social trends such that, to this reader at least, the point is often lost. It is true that to touch one strand in the spider's web is to cause reverberations in all the connected strands, but in describing them, sometimes Taylor seems to be losing track of the spider herself.

At the other pole from Taylor's approach is a shorter book of fewer than two hundred pages by Robert Coles, *The Secular Mind*, which takes a limited historical view. The book begins with the situation in America in 1900 but is largely a personal description of Coles's own experiences as a psychiatrist and student, linked with many references to modern literature. Coles's style is soothing and informed, and in the end, he is sympathetic to the new mood of secularism, finding in its contemporary form a kind of "sanctity," a new mode of being in which religious belief, having been removed from the psyche of modern humanity, is supplanted by involvement in the issues of day-to-day living.[6] For Coles, the absence of a sense of a system of moral reference that exists independently of one's own personal opinion presents no actual problem.

Taylor's treatment of secularism treats it as an untoward event in intellectual history, and Coles's approach seems mostly passive, as though it presents a not-necessarily serious dilemma in psychological adjustment. The power of the secular movement is thus illustrated as an intellectual and psychological phenomenon, which shows the extent of its contemporary influence. But secularism is rightly termed to have *power*, in the sense that it, however vacuous as a general philosophy, has become instantiated in the American state. It is a form of irreligious incarnation, one might say, as secularism propounds a state policy in which religious doctrine is explicitly denied a role or a place. Even a scent or a whiff of religiosity in a civic debate is enough to put the hounds of un-heaven on the case to chase away the offending presence of faith in almighty God.

The same holds true in the European context, for in Western Europe there has been what its critics describe as a hollowing out of its culture, namely the Christian religious sensibility that basically formed European culture. Two prominent critics of what we term the secularization of Europe coincidentally, in two successive days, presented detailed account and criticism of the hollowing out—Cardinal Ratzinger and Professor Marcelo Pera. Ratzinger, later Pope Benedict XVI, gave his critique from a religious point of view (obviously), decrying the relativism that has evacuated any sense of an objective moral order. Pera, a humanist philosopher, was particularly concerned that the decline of the influence of Europe's Christian heritage was leaving Europe vulnerable to threats from Islamic expansion. In his foreword to the volume that reprinted Ratzinger's and Pera's presentations, American historian George Weigle made the connection between the de-Christianization of Europe and the decline in its population caused by the failure to reproduce.[7]

Science and the State

The final effect of modern science has to do with the state: that modern science has provided the technological means for states to increase their power should not go without saying; nonetheless, it is an effect that is not often noted. Modern science enables the expansion of the physical power of the state into new and unanticipated realms, the result of which is, in effect, a new kind of state with reserves of power unthought-of before the middle of the last century. Scientific technology gives the state the physical ability to more easily conduct its affairs, gather intelligence, and project its power (including the application of military force by means of advanced weaponry, all of which can be applied to its own citizenry as well as foreign enemies). Consider the speed and nature of communication and information gathering for just one example (which leaves aside such things as the development of weapons such as the machine gun, nuclear arms, and drones).

In the days of ancient empires, a revolt on the Rhine could not come to the notice of Emperor Augustus for many days or even weeks, as the fastest means of communication was by horseback or ships powered by sails. The emperor had then to rely on reports, verbal or written forwarded by messenger, from persons at the scene who were not present to him and could not be interrogated for significant details. In such circumstances, the emperor had serious difficulty in deciding what actions to take, usually sending out a military force under the direction of a leader whose judgment he could trust, since the emperor himself would not be onsite and would thus be denied knowledge of what the circumstances were. Today, a modern tyrant such as Assad in Syria does not operate under such restrictions. Since the invention of the telegraph (first used during the Civil War) and subsequently with the invention of radio, television, and satellite transmission, the tyrant has instantaneous knowledge of a revolt in the city of Homs—or of any other crisis. He not only is notified by report, but, with up-to-date communications technologies, he can interrogate the messengers and can actually see for himself the circumstances of the crisis. (The same ability also applies to democratically elected leadership, of course.)

But the effect of modern science on the empowerment of the state goes beyond providing the technological means for a degree of control unavailable before in human history, for it derives from the fact, noted above, that modern science is in the process of losing its status as the emblem/symbol of Enlightenment reason. It may seem contradictory, but it is an observable sociological fact that scientific research results are first of all being massively supported by means of government grants (through the NSF, NIH, Defense Department, etc.) but then that research results are being politicized, that is are being utilized or suppressed depending on whether they support or contravene government policies.[8] Scientific research concerning subjects with wide social consequences is expensive and because it examines subjects of high complexity, its results are often debatable. The net result is that government policies that are directed by ideological or political motives are given the color of a false rationality by justifying them as based on scientific research.

In Summary

The book is divided into seventeen chapters and five major parts. There has been a major change, rich and strange, in American culture, whose consequences are not yet fully revealed but which, given the nature of human history, cannot all be positive. Previously, up until the middle of the twentieth century, the Protestant Christian religion in multiple forms provided the focus of cultural unity, which manifestly, it no longer does. This has had an extraordinary effect on American culture and so requires an account. Explaining the origins of the multiple forms of the Protestant religion in America, how the American religion declined in the case of a major university originally founded under religious auspices, and what recent responses the decline has provoked is done in Part I.

That as religion has declined in influence, science has risen, there can be no doubt; that the rise of science is solely responsible for the secularization of American culture, there is serious doubt. The present state of science within the culture has changed from what it once was: effectively, the symbol and guarantor of human rationality and its ability to reorder the world according to human wants and desires. For the general and precisely philosophical understanding of science has changed from a belief in its inerrancy, due to its method and neutrality, to a recognition that the thought processes involved in scientific discovery are no different from those of other fields of human endeavor—relying, that is, on prior assumptions, community identity, and hope. The actual state of science as it presently exists in American culture, including the place of what can be called "postmodern physics" is analyzed in Part II.

Despite the decline in confidence that science tells us final truths and that technological progress brought about as a result of scientific research is an unalloyed "good thing," some scientists have believed that their particular fields can provide a basis for the culture. The work of three American scientists prominent in their respective fields—Weinberg in particle physics, Wilson in evolutionary biology, and Milgram in social psychology—are examined in this regard in Part III.

The overall result at this point in the exposition inclines the observer to a pessimistic outlook, as we consider the decline in confidence in science and the current state of American religious belief as it loses any sense of unity and breaks up into various conflicting fragments. Most disturbing, perhaps, is the failure of America's own particular school of philosophy, for Pragmatism currently cannot provide any overall explanation of the relations between religion, science, and culture, even though its founders, Peirce and James, had made serious efforts to do so. Most recently, pragmatic philosophy has refused, as it were, to provide an intellectual basis for social change, as Dewey had for the Progressive era. The description of these failures is the burden of Part IV.

There is an implied disconnect between scientific reason and the aims of the secular state, which are political and ideological in nature. But to make the case that there is a contradiction between scientific reason and the ever-expanding claim of state power, it is necessary to separate scientific reason expressed in theories and laws from their technological application; for in our culture, technology has overwhelmed the appreciation of science as a theoretical enterprise—for example, as the chief provider of master narratives. Meanwhile, religious belief persists in American culture, albeit not in the cultural dominance of any organized denomination, but in the form of a general belief in the One God. The new secular state has arisen and is growing in power and implied menace before our very eyes; it ignores, when necessary to achieve its political goals, both the results of scientific research and the moral commands of religious faith. The expansion of its own power now appears to be the *raison d'etre* of the secular state. All this is described in Part V.

In conclusion, given the rise of a new secular state and all that has been accounted for beforehand in the mutual relations of science, religion, and secularism, the final point is nonetheless optimistic, for cultural disharmony can assure that the new secular state never achieves the dominance over culture and political life it inherently desires.

Notes

1. See Hitchens, Christopher, *God Is Not Great: How Religion Poisons Everything*; New York, Hatchett Book Group, 2007.
2. Taylor, *A Secular Age*; Cambridge, Harvard U. Press, 2007.
3. E. O. Wilson, *On Human Nature*; Cambridge, Harvard U. Press, 1978; 201–209.
4. Taylor, *Secular Age*; 99.
5. Ibid.
6. Robert Coles, *The Secular Mind*; Princeton, Princeton U. Press, 1999; 189.
7. Pera, Marcello, Joseph Ratzinger (trans. M. Moore), *Without Roots*; New York, Basic Books, 2006; Weigle, reference, viii, ix.
8. Greenberg, Daniel, *Science, Money and Politics*; Chicago, U. of Chicago Press, 2001.

Part I

Origin and Decline of the American Religion

Chapter 2

America, Inventor of Religions

Inventive America, Golden Plates, and Cyclotrons

America's inventive ingenuity is often praised; however, the examples cited are almost always technological marvels such as television, the steamboat, the cotton gin, the telegraph, the airplane, the electronic chip, and so on. Edison is held up as a great example of the American success story, or Steve Jobs—men who, in an entrepreneurial fashion, devised many separate inventions and brought them before the American public in a way that made positive and important changes in daily life. But extending beyond the inventions themselves, and reflecting on the business side by which these inventions were developed and brought to market, makes us aware of another field of American inventiveness, namely in manufacturing techniques, the primary example being Henry Ford, who developed the assembly line for the manufacture of his inexpensive and popular automobiles. Beyond this, another aspect becomes apparent: namely, the invention of new types of business organization, as exemplified in the vertical organization of the oil business, from wellhead to refinery to gasoline station, by Rockefeller; and of the personal computer business by Gates, from operating system to internet to applications to hardware. The "grandfather" of this general kind of American inventiveness is Benjamin Franklin, a go-getter who invented the Franklin stove, bifocals, and the post office and who developed his own philosophy of go-getting—not to mention his adventures with the kite. Franklin not only did experiments with electricity, but he invented a new *theory* of electricity; namely, that it came in positive and negatives forms. Benjamin Franklin was a theorist as well as an inventor.

But the concept of American inventiveness needs to be extended from technological devices and business organization to an appreciation in general of the American character as inventive in itself, a trait which was made necessary by the fact that Americans had to fill up a (mostly) empty continent, importing traditional forms of life from their countries of origin, inventing new forms of social organization, and adjusting for a continual rush of new circumstances, cultural as well as geographic. From there, we need to proceed to an understanding of how the American spirit of invention applies to areas not generally included: governance, philosophy, and, most notably, religion. As symbols of respectively technological and religious inventiveness, we can note that Lawrence in 1930 invented the *cyclotron*, which was the first "atom smasher," a necessary device for revealing the internal structure of atomic nuclei and the prototype for the CERN Supercooled Supercollider in Switzerland, now searching for the ultimate atomic particle. A century earlier, Joseph Smith discovered golden plates whose location was revealed to him by an angel, thus "inventing" Mormonism, which critic Harold Bloom asserts is the most typical of America's invented religions.[1] American inventiveness is exemplified not only in the plethora of technological innovations the nation has produced but also in its many native-born religious denominations.

Sources of Religious Invention

America has invented as many separate religions as it has major technological inventions, and the list of native American religions—that is, those that were invented on American soil in the context of American culture—is impressive.[2] These new religions have thrived and grown in numbers so that several of them count their adherents in the hundreds of thousands—and even millions!—and have sent their missionaries beyond the borders of the United States. For most of the new American religions, there is a common and explicit source, namely the King James version of the Bible. It has become, as a matter of historical circumstance, the talismanic item on which certain forms of Protestantism base their doctrines, as the replacement for Catholic tradition and papal authority at the time of the Reformation. Here, in full, is the 23rd Psalm as it appears in the King James version, its familiarity not diminishing its beauty and power to this day:

> The Lord is my shepherd, I shall not want.
> He maketh me to lie down in green pastures: he leadeth me beside the still water.
> He restoreth my soul: he leadeth me in the path of righteousness for his name's sake.

> Yea, though I walk through the valley of the shadow of death, I will fear no evil:
> for thou art with me; thy rod and thy staff they comfort me.
> Thou preparest a table before me in the presence of mine enemies;
> thou anointest my head with oil; my cup runneth over.
> Surely goodness and mercy shall follow me all the days of my life:
> and I will dwell in the house of the Lord forever.

But the utter reliance on one source, the Bible, for Christian doctrine and practice, combined with the Protestant view declared by Luther, that all Christian believers can read the Bible for themselves without commentary or tradition to filter its true message—the truths imparted by the Holy Spirit speaking to the believer in her heart—has led to a plethora of interpretations of doctrine, some of them quite far from any traditional understanding of the Christian faith. In this way, traditional Christian doctrines have been combined with personal beliefs unique to the reader of the Bible and representative of certain cultural currents as well, being integrated with varying degrees of coherence into traditional Christian doctrine, as exemplified in the Nicene Creed. Among the radical departures from traditional doctrine, the most apparent concerns the Trinity, the idea that there are three persons in one God, as expressed in the Nicene and "Apostle's" creeds. This is a difficult—which is as it were *impenetrable*—doctrine; and, in the past, various Christians have denied it because of its apparent contradiction of basic human reason (e.g., the Bishop Arius in the fifth century and Isaac Newton in the seventeenth).

The doctrine of the Trinity is connected inherently with the status of Jesus Christ, for denial of the Trinitarian doctrine implies logically that Jesus is not coequal with God the father, and thus denial of the Trinity means reduction if not outright denial of Jesus's divine status. This is apparent in some of the new American religions, for Unitarians, Jehovah's Witnesses, and Mormons all reject the doctrine of the Trinity. Such new religions may be termed "post-Trinitarian" religions, a term that also applies to Islam, in whose doctrine Muhammad is the last and greatest of the prophets, but Jesus is the second greatest. (Muslim theology emphatically denies the doctrine of the Trinity, arguing that it is simply polytheism.)

Many of the new American religions not only dispensed with uncomfortable elements of the traditional creeds of Christianity, they also integrated aspects of American culture into their doctrines. One of the most profound and self-sustaining of these is the sense of "self-reliance," as Emerson, an early "prophet" of the American religion put it, now understood as self-fulfillment and self-empowerment (and by many other contemporary terms). The most important aspect of the emphasis on individual self-fulfillment is readily apparent in the emphasis on personal experience of the divine, finding that God resides within the self, which has become a virtual trademark of some recent neo-Christian religions, especially among so-called *mega churches*.

But such a doctrine is not so much a radical departure as a peculiar emphasis on the tradition of mysticism in Christian doctrine and practice, readily apparent in some of the writings of St. Paul, as when he says he reached "the seventh heaven." Accounts of ecstasies and voices experienced by holy people have always accompanied the Christian tradition, including such figures as Bernard of Clairvaux, Francis of Assisi, Tauler, Julian of Norwich, Teresa of Avila, Joan of Arc, George Fox, and Thomas Merton. However, reports of mystical experiences have often been treated with suspicion by ecclesiastical authorities, precisely because the mystics frequently impart some new revelation that may contradict, in some way, received doctrine. Such events have often taken place in the development of new American religions, including the revelations of Joseph Smith and Mother Annie Lee. More commonly, however, the emphasis on the self has reinforced the innate desire of religious believers to have personal experience of the divine—in ecstasies, glossolalia, or simply and quietly within themselves. This has been the case especially when believers feel that the practices of their denominations emphasize the formal aspect of religion rather than the experiential; then, believers will reject the "cold" formalisms of doctrine in favor of the warm, direct experiences of the divine, which explains the appeal of some of the new religious movements in contemporary America.

Two things account for the profusion of new religions: First, American inventiveness, as has been pointed out. There is another cause, namely the fact that in the expansion of American pioneers westward, the new settlers had to set up their own institutions, such as law courts and schools. They also had to provide for the innately human need for religious worship, for there was no national Church, as in England or Italy, and thus no omnipresent clergy and ready-made houses of worship; for there remains a need for explaining such a flowering and such a powerful expression of religious worship, an underlying cause universal not only in American culture but universally in human nature. Rejecting Bloom's thesis that all the new religions in America should be understood as a form of *gnosticism*, I am looking at a number of separate causes including religious doctrines, texts, nonreligious cultural movements, and historical circumstances. There is also implied in the flowering of "the American religion" in all its multitudinous forms, a powerful need common to humanity to seek a transcendent explanation or ground of lived, existential, day-to-day reality.

America's New Religions

In an account of the major technological advances made in America, we would list the names of the inventors and provide a brief description of their inventions. So also with regard to America's invented religions, here is a listing of the major, distinct denominations that have arisen in the United States of America.

Jehovah's Witnesses

The presence of Jehovah's Witnesses is common in the northeast United States, where many unsuspecting people have met the believers at their front doors as the Witnesses patiently but persistently hand out their literature, including the small, broad sheet "Witness" or their famous "blue book" explaining their view of the theory of evolution. (Steven Jay Gould, evolution's best-known exponent in the late twentieth century, claimed that the "blue book" was the most common source of most people's knowledge of the theory.) As explained below, the Jehovah's Witness religion stems from the Millerite phenomenon in the middle of the nineteenth century and thus shares with Seventh-day Adventists a highly critical attitude toward contemporary American culture, opposing its materialism and its ignorance of the coming Judgment delineated in Revelation. As implied in its name, "Jehovah" being the name given to the first person of the Christian trinity, the religion tends to denigrate the status of Jesus, claiming that he is not to be identified with the Father, but rather that he is the creation, not the coequal, of God. Thus, Witness doctrine claims that Jesus and Satan are brothers, equals in effect, who are at war with one another, explaining the conflict between good and evil seen on this earth. This contradiction of traditional Christian doctrine is reinforced by the Witness's peculiar translation of the first line of the Gospel of St. John, which proclaims that "the Word was with God and the Word was God." The Witness Bible translates that portion as "the Word was with God, and the Word was *a god*" (emphasis supplied).

Mormons (Latter Day Saints or "LDS")

Upstate New York in the early nineteenth century was the scene of many religious movements wherein numerous preachers—often unattached to any formal religious organization—gave sermons to crowds of earnest seekers after the Word; so many such movements abounded that that part of New York was called "the burned-over district." And while most of this preaching was based on the King James Bible, and thus the theology more or less Protestant, some of it was unique and only indirectly based on traditional Christian doctrine. Joseph Smith was such a preacher and is the prophet who founded the Church of Latter Day Saints, also called Mormons. His doctrine had elements that widely copied the Christian teaching but recast them in a peculiar form. Thus while worship of Jesus was central, more attention was paid to the Trinitarian Father figure, and Jesus was reduced to less than identity with him. But Jesus was also said to have visited the American continents, North and South, for the Mormon scripture is a copy of the Old Testament story of the ancient Israelites transferred to unknown tribes in the Americas. Other new doctrines proliferate in the LDS religion, but its doctrine is probably less important than its current presence in the United States, where the number of Mormons has grown tremendously and where they have attained significant influence in business and government and are widely known for their personal integrity and competence. (As of this writing, a Mormon, Mitt Romney, is the likely Republican candidate for the presidency of the United States, and several US senators are also Mormons.)

Christian Science

Mary Baker Eddy was a sickly woman who lived in the early nineteenth century and who was religious in the general Protestant manner. She was impressed with the accounts of Jesus's many healing miracles in the four Gospels and eventually found strength—spiritual, psychological, and physical—from meditating on these scenes and others in the Bible. Her book, *Science and Health With Key to the Scriptures*, is the founding document of the Christian Science religion, which has a comparatively small number of adherents but which is nonetheless influential. The Mother Church is an outstanding architectural expression of the faith on a plaza in downtown Boston that is the frequent destination of pilgrims and sightseers. *The Christian Science Monitor*, now a weekly, has been a model of journalistic integrity and competence since 1908 and is the best-known representative of the religion to the American public. Christian Science is known for its rejection of modern scientific medicine and its reliance on prayer instead. Sadly, however, there have been incidences of children and others dying from causes that modern medicine could have effectively aided.

Southern Baptists

The ultimate origin of the Southern Baptist religion is obscure, although the Baptists in their earliest form on the American continent are seen in New England, as a transfer from European Anabaptists whose practice was to delay baptism until the age of reason, making it a rational choice in opposition to infant baptism. The Baptist tradition was stubbornly independent of collusion with any form of civil authority, and thus had to escape from the oppressive culture of Massachusetts in the seventeenth century, which was dominated by Puritanism. The Baptists are proud of their insistence on religious freedom in the United States, having influenced Thomas Jefferson's declaration that a "wall of separation" should exist between government and religion in America. In the deep south of the United States, the Southern Baptist religion has grown and thrived as a main element of the "Bible Belt," to the point where the Southern Baptists dominate the local culture. The Southern Baptists are notable for two reasons in contemporary terms: first, for their adherence to traditional Protestant beliefs regarding the presence of sin and necessity of faith in seeking salvation, but also for their direct political influence as "value voters." The founder and moving spirit of the "moral majority" movement in the latter part of the twentieth century was the prominent and controversial Baptist preacher, Jerry Falwell.

Fundamentalism

"Fundamentalist" has become a term of derogation, referring to how religious beliefs in general can become degraded to rigidity and fanaticism. Fundamentalism, however, is an actual movement in traditional Protestantism that was developed in the early twentieth century.[3] The need for an explicit, minimum standard for traditional Christian belief was recognized as a direct result of the weakening of traditional Protestant belief brought about by two causes that arose in force in the middle of the nineteenth century. First was the study of the Bible as an ancient text comparable to other texts produced in the ancient world, rather than as the revelation of the Word of God. Second, the intense interest and controversy brought about by the announcement of Darwin's theory of evolution, which seemed to put science in direct opposition to faith. In response, a group of Protestant laymen and ministers defended fundamental theses assumed to be essential to the Christian faith—a kind of creed that established a solid bulwark against the seepage of modern ideas that was undermining (and continues to undermine) the Christian faith.

Pentecostalism, Assembly of God

The long decline of membership of the mainline Protestant churches, when it is commented on, is usually compared to the increase in evangelical or Pentecostal churches, including, for example, the Baptists.[4] Perhaps the most representative of this group of denominations is the Assembly of God church because it emphasizes what the mainline churches seem to have lost: namely, an emotional connection with the divine realities that Christian doctrine identifies. As mainline churches have increasingly devoted themselves to social-justice issues, their practice has seemingly devolved away from emphasis on the personal experience of Christian living (as when one pious Protestant lady asked the author, "Does Jesus sit on your shoulder?"). The Pentecostal movement became a recognizably separate denomination in California in the early twentieth century, its doctrine and practice supported by reference to the Pentecost described in the early chapters of the Acts of the Apostles, the book of the New Testament that traditionally is placed after the four Gospels. Pentecost, named from an ancient Jewish celebration, describes the point at which the third person of the Trinity, the Holy Spirit, entered the room where Jesus's apostles had met (along with Jesus's mother), which resulted in the enspiriting of the followers with extraordinary gifts (as mentioned by St. Paul in his various letters) including glossolalia, prophecy, ability to interpret scripture, and mystical ecstasies.[5] Pentecostalism looks to renew these experiences as an essential part of the Christian experience and, in the form of Assembly of God churches, has shown tremendous growth. The theology of Pentecostal churches retains a largely Protestant flavor yet does not fall into the predestinationism of Calvinist theology seen in the Congregational and Presbyterian denominations that took root on the American continent in the seventeenth and eighteenth centuries.

Besides Assembly of God, another manifestation of Pentecostalism is the Four Square Gospel Church, founded in California by Aimee Semple McPherson, who is usually and unfairly described in popular culture as just another interesting personality from the West Coast, whereas she is more properly compared to Mary Baker Eddy and Mother Annie Lee as an example of a woman founding a major American religious denomination.

Seventh-day Adventist

It might seem that a self-proclaimed prophet who made a spectacular prediction based on an interpretation of certain biblical passages—and then had the prediction spectacularly fail—would indeed be a prophet without honor, leaving no institutional descendents behind him. In the middle of the nineteenth century, William Miller announced to his followers the exact time in 1843 when Jesus as Lord would return to earth to claim his eternal kingdom, a prediction that failed not just once but twice (Miller corrected his original calculation several times). Nonetheless, Miller's followers did not disappear into the general Protestant populace but effectively formed two new American religions: Seventh-day Adventists and Jehovah's Witnesses. What accounted for this? Both religions, as they currently exist, spring from Miller's notion that the current earth and the American nation in particular were inherently sinful, and that mankind lives in a sinful time that has earned God's judgment. Many people who saw the destruction caused by two world wars, the Great Depression, and continual international conflicts, plus materialism and moral decay in American society, easily accepted the apocalyptic aspect of Miller's inspiration. The Seventh-day Adventists made their peace with Miller's false prediction and kept up the sense of moral decay and godly judgment. Adventists have made physical health a special cause, running hospitals and encouraging medical science; also, they continue to refuse the usual and expected forms of reverence to the United States of America (declining to salute the flag) and keep the Jewish Sabbath day of Saturday.

Black Muslim

It may be said that the one good thing that American slave owners gave to the African slaves was the Christian religion. It became perhaps the only help available at a time of great distress and peril for the slaves and provided them with the biblical narrative of the

Israelites' escape from slavery in Egypt. In return, the slave culture gave American Christianity its soul force and its plaintive and moving music. Nonetheless, many descendents of the slaves identify the Christian religion with slavery itself, even though it was Protestant ministers who, in the 1960s, led the Civil Rights crusade. This rejection is most flagrantly exhibited in the Black Muslim religion, which, founded by Wali Fard in the 1920s, came to prominence in the 1960s, promising violent resistance rather than nonviolence as the means for black Americans to acquire their full rights. The religion itself has broken apart but retains its influence and generally presents a dynamic but angry face to what it sees as the racism endemic in white culture. The best-known example of this is the use of Muslim or African names: Cassius Clay becomes Muhammad Ali, Lew Alcindor becomes Kareem Abdul Jabaar, Malcolm Little becomes Malcolm X. The rigidity of the Black Muslim religion and the intense level of indoctrination—which includes anti-Semitism as well as a general accusation against the perfidy of American institutions—is frankly frightening, yet the basis of the appeal of the religion to many African Americans is largely positive. In an inner-city environment, joining the Muslims is a means of literally extending the life expectancy of young men and women, who will find a support group as they leave the bad habits of the city streets, moving away from drugs, violence, crime, and general dissipation, and develop habits of responsibility, work, and family life, all of which are emphasized as aspects of the Black Muslim experience.

Dianetics

While the origins of most religions are obscured to some degree by the passage of time, that of Dianetics or Scientology is not. It was founded by L. Ron Hubbard in the 1950s and is based on neither the Bible nor any revelation but on an understanding of modern scientific psychology. Dianetics promises good mental health based on contact with the inner self by utilizing certain mental exercises mediated by use of the *orgone box*. In the early 1950s, Hubbard explained the orgone box and its effectiveness in articles this author remembers reading as a teenager in *Astounding Science Fiction*. Dianetics stands out as an attempt to put religion on a rational, scientific basis, although as it has developed a "back story" involving super-intelligent aliens, it seems to have devolved into its origins in science fiction. Dianetics has adherents among some celebrities, including Hollywood actors, and is presently being promoted through the distribution of books and CDs.

Unitarianism, Universalism

The tendency to question the doctrine of the Trinity on rational grounds reaches its maximum point in the Unitarian religion, a denomination that is not large but is very influential. The early days of colonial America in the northeast (and Massachusetts particularly) were dominated by the heavy-handed puritanism of the Congregational church. (Most of the traditional white clapboard churches that are reflective of New England character are Congregational churches.) Reaction against the Calvinist doctrine of eternal damnation combined with predestination resulted in the development of new, more liberal doctrines among educated believers, especially those trained at Harvard College, whose first mission was to be a Congregational seminary. This development implied a separation between the beliefs of the churchgoing believers and those of the more educated. Unitarianism's rejection of the Trinitarian doctrine as expressed in the Nicene Creed has historical resonances, for it was a doctrine first propounded in the fifth century by the Bishop Arius; and over time, many Christian believers have found cause to wonder about the confrontation of the Trinitarian doctrine with the simple human logic that three cannot equal one. In Unitarianism, the Christian message is understood to emphasize particularly the directive to do good to those less fortunate; the Trinity and the Incarnation are both rejected, and Jesus is reduced to the status of a good man who is an ethical teacher and an exemplar but not divine.

The doctrine of eternal damnation and concentration on mankind's innate depravity, so much emphasized in the regular practice of the Puritan religion and exposed in Hawthorne's *The Scarlet Letter*, led some ministers and believers to reject these doctrines altogether. A new, more liberal theology was developed—spread by prophets like John Murray—that took root away from the city of Boston in the hinterlands of Massachusetts farms and villages; namely, that no one was condemned to damnation since Christ's message of eternal salvation applied to all men without distinction. This resulted in a purgatorial doctrine in which some bad people were punished after death but not eternally. By the middle of the twentieth century, Unitarianism and Universalism, both actual denominations with membership lists and churches, had lost their connection with traditional Christian doctrine, and the two denominations were formally combined in 1966. The Unitarian Universalist denomination, as it is known, has churches throughout New England and other parts of the country and supports the Beacon Press in Boston, which publishes highly regarded volumes on culture, history, and religion that are usually of a liberal bent.

New Age, Wicca, Gaia

There have been other recent religious developments in America, but they seem more reflective of trends in the general culture combined with the normal human desire to find connectedness with higher powers, nature, or mankind than to form actual religions. It is difficult to know where to draw the line, since each of these new developments are so diffuse, lacking an authority structure and canonical scripture. Also, although it is entirely possible to have a religion without reference to the biblical God, or to any god for that matter, most of the tendency in these new movements is toward human beings in themselves in terms of self-development, self-fulfillment, mental balance, and physical health. Religion in some fashion would seem to mean referencing human behavior and physical reality to something or someone outside themselves, and if so, many of these new movements do not qualify as religions.

* * * * * *

This list does not include the many separate houses of Protestant worship that are most often Bible based and that are expressions of Reform Protestantism emanating from Calvinist roots. Nor does it include the revised forms of religious organizations which were originally imported from Europe but which have undergone significant "Americanization." Judaism has been extended into no less than four recognizable forms: Orthodox, Conservative, Reform, and Reconstructionist, each with their own seminaries and congregations. The Pietist strain of Protestantism imported from northern Europe has flowered into the Quakers, Shakers, Amish, and other similar offshoot religions. Catholicism, while not differentiated in terms of doctrine and organization from the Church in Rome, has nonetheless become modified into a typically American version emphasizing practicality and personal liberty, which in turn has had an effect on Catholic doctrine and practice regarding religious freedom.[6] An interesting point of speculation in this regard concerns Islam: will an American version of Islam come to pass as a result of Muslims living, after several generations, in America? And will America be as successful in integrating the Islamic faith into the American context as it has been with other faiths?

The Episcopal Church Divided

The Church of England is the established church of England and is the compromise between two opposed religious traditions, Roman Catholicism and Reform Protestantism. When Henry VIII dissolved the monasteries, rejected the authority of the pope, and made himself the head of the Church in England, he did not largely act to change the liturgy or the doctrine of the Catholic Church; those were left in place. It is for this reason that Henry is called a "conservative" reformer, especially in contrast to the Protestant Reform movement that later had influence in England, for it had become a safe harbor for Reformers subject to prosecution and death by (mostly) Catholic authorities on the continent. It was Henry's daughter Elizabeth who attempted a compromise between the Catholic and the Reform Protestant traditions, a compromise that has left the Church of England subject to tensions and divisions and that led to outright warfare during the civil war in the seventeenth century. Such a division is currently taking place in the Church of England, and even though it has not resulted in civil war, it has forced perhaps the worst crisis in the history of the Anglican Church. That crisis is the result of events that originate, and are currently taking place, in America.

As a result of the American War of Independence, the Church of England in America changed its name—but not its identity—from the Anglican Church to the Episcopal Church. The Episcopal Church remains part of the Church of England, which is now a worldwide church, and remains under the overall authority of the chief bishop of the Church of England, the archbishop of Canterbury; however his control is far less effective than that of the Roman Catholic pope. The influence of the Episcopal Church in America is reflected in the fact that many prominent and influential Americans are and have been members—including, as the primary example, George Washington. The national cathedral of the United States, located in Washington DC, is an Episcopal church, albeit ceremonies from virtually all religions are held there. The authority structure of the Church, reflecting its history and origins, does not follow the Roman model but is collegial (i.e., individual bishops and dioceses have a great deal of latitude in both doctrine and practice).

The current crisis happened because ideals developed in the secular culture have invaded (or influenced, in a more neutral expression) many Episcopal churches and dioceses in the United States. The immediate issue has come down to ordination of women in the Episcopal priesthood and consecration as bishops. Interestingly, the argument for female ordination has not usually been made in biblical terms (e.g., that, given Jesus's comparatively accepting attitude toward women, if first-century mores had allowed it, he would have included women among his apostles, such as Mary Magdalene); rather, the argument has been made in terms of inclusiveness and toleration, ideals that are worthy if secular. The issue has been exacerbated by being extended to include homosexuals and transvestites, and so the ordination of gay and lesbian priests has followed, climaxing in the state of New Hampshire by the presence of an Episcopal bishop living in an active homosexual relationship.[7] In this conflict between the secular ideal of toleration and the Christian doctrine that certain types of sexual behavior are sinful, the ethic of toleration seems to negate the concept of sexual sin and sometimes, it seems, of personal sin altogether.

The results have been a schism between the liberal activists who now run the Episcopal Church and the traditionalists, which in turn has resulted in a gulf of separation such that there are now parallel universes of Episcopalianism in America. Disaffection among the traditionalists has led some of them to develop formal identification with Episcopal bishops from Africa. There have arisen a number of ugly court cases over which group owns the assets of certain Episcopal churches, each side claiming the other has moved illicitly to take over ownership of church buildings and financial holdings.[8] The schism that has arisen in America is having worldwide consequences for the Church of England; the issue has been the subject of debate and policy changes in the Anglican Church and is the chief issue faced by the Archbishop of Canterbury, and even the current queen, Elizabeth II, since the English monarch is still officially the head of the Church of England. As is usual when storms afflict the Episcopal Church, many of the dissidents have become Roman Catholics, and Pope Benedict XVI has put in effect a means by which whole Anglican parishes and dioceses can join the Catholic Church while retaining elements of Anglican liturgy and practice.[9]

The liberal Episcopal Church is greatly concerned with issues of equality and social justice, as the Church has been historically in the United States; however, its reliance on traditional Christian doctrine is eroding, and its membership has declined to about three million—fewer, it is said, than the number of Muslims in America. The Episcopal Church in America, it now seems possible to predict, may go the way of the Unitarian Universalist Church and ultimately lose its Christian identity altogether. But if so, this will not have been the result of the influence of scientific atheism or of scientific materialism but of a self-redefinition of the Episcopal doctrines and practice by a significant number of its own adherents.

Notes

1. Harold Bloom, *The American Religion*; New York, Simon and Schuster, 1992; 80.
2. Patrick Allitt, *Religion in America Since 1945: A History.* New York, Columbia U. Press, 2003; 7.
3. Ibid., 181.
4. See "News from the National Council of Churches," 2011 Yearbook of American and Canadian Churches regarding decline of Main Line; http://www.ncccusa.org/new/2011.l.
5. On glossalalia, First Corinthians, 12–14.
6. The text of "Dignitatis Humanae," promulgated on Dec. 7, 1965, is available on the Vatican website; http://www.Vatican.va/archive/hist_councils/11_vaticar.
7. Bishop Gene Robinson tells his own story in *God Believes in Love: Straight Talk About Gay Marriage*; New York, Knopf, 2012.
8. See George Conger, "No break in pace of Episcopal Church lawsuits"; http://geoconger.wordpress.com/2010/08/09 (accessed 12-8-2012).
9. "Pope formally creates group to attract dissident Episcopalians," Episcopalian News Service staff, Jan. 1, 2012; http://episcopaldigitalnetwork.com/ens/2012/01/03 (accessed 12-8-2012).

Chapter 3

How Tufts University Lost Its Religion

As extraordinary as is the appearance of so many separate religious faiths originating in America is the contrary phenomenon of the general decline its religious belief. The decline of religious belief in the United States is evident on many fronts—for example, in the decline of membership of mainline Protestant churches such as the Episcopal Church or in social-science surveys of religious attitudes, which reveal, for instance, that while many people call themselves "spiritual," they do not belong to any organized religion.[1] One of the most significant arenas where the decline of American religious identity is apparent—as well as being possibly the most influential—is in America's major colleges and universities. One example of a major institution will suffice as an indicator of the whole.

Tufts University Founded as a Protestant Institution

Tufts University lies north of Boston, one of the several colleges and universities that give the Boston area its reputation as a center of higher learning. Tufts is a true university, its highly regarded medical and dental schools reside in Boston, and its equally regarded veterinary school resides in North Grafton, a town just outside the Boston commuting area. Tufts University supports a chaplaincy system for the students, manned by a senior chaplain and three associates—one Protestant, one Catholic, and one Jewish. The main campus is on the line demarcating two blue-collar suburb cities, Medford and Somerville, which contain the undergraduate schools of Liberal Arts and Engineering; the graduate school, which grants masters and doctorate degrees in a number of fields; a separate department that grants degrees in education; plus the Fletcher School of Law and Diplomacy, whose best-known graduate is the late senator Daniel Patrick Moynihan. Tufts University was founded as a liberal arts college in lieu, it appears, of a Universalist seminary, since one of the principle aims of the founding of Tufts College (it was not named a university until later) was to educate Universalist clergy. Architecturally, the campus is an eclectic mix of buildings—some over a hundred years old, revealing a heavy Victorian sensibility with columns and old dark brick; some modernistic and made of plate glass and new brick; a campus center with low-lying red roofs, making it look vaguely oriental; and large, imposing buildings, built in the forties and fifties, which house the Engineering and Chemistry departments. The eclecticism of the architecture can be taken to represent the various factors that have formed Tufts University over the years, beginning as a one-room college in Ballou Hall, which still stands and is used as the main administrative building. The latest building to be erected is a seven-story parking garage, the top floor of which now houses all of the student-services offices, the current emphasis seemingly less on the sacrifices necessary for the attainment of higher learning than on providing a multitude of services for the students who are now seen not as acolytes but as consumers. (But then, Tufts charges approximately fifty thousand dollars a year, and the parents who pony up the money must be assured that they are getting something for such a high price.)

On the highest hill sits one of the oldest buildings on campus, the Goddard Chapel, which is quite distinctive; it is a complexly shaped building, constructed of large, irregular pieces of fieldstone, featuring steep slate roofs and a tall spire in addition to a square bell tower that is capped by a squat-looking cross. Inside, the chapel is dark, painted in blue and brown, the colors of Tufts; and with brown wood predominating, it is adorned with plaques and busts from the nineteenth and early twentieth centuries. Sunlight enters through stained-glass windows that feature thematic figures, such as a sower casting seed; and, over the altar area, a very large, semicircular window portrays an ancient-looking man dressed in robes and holding a sword and a large book. The odd thing is that there is nothing distinctively religious about the man—it could be Aristotle just as well as St. Peter; and this lack of the obviously religious is a feature of the decorations and other windows of the chapel (with the possible exception of the semicircular window at the rear of the chapel, which might be a portrayal of St. John the Evangelist). This lack is not accidental, nor is it the result of recent changes in attitude toward religion, for the chapel was built for—and the university founded by—Universalists, the Protestant denomination that eschewed what it termed "sectarianism" and which, in revolt against the stark predestinationism of the ambient Puritan Calvinism, believed that all men, without exception, were saved by the sacrifice of Jesus Christ (see Chapter 2). This Universalist version of the Christian gospel, which eschewed conversion experiences and claims to religious exceptionalism, was preached in colonial times by a man named John Murray, for whom a Universalist church is named in Attleboro, Massachusetts. Murray's evangelization raised severe opposition at the time, and the Tufts University libraries stock a large number of books (as well microfilm volumes) that document the debates at the time, during the late eighteenth and early nineteenth centuries.

The doctrine of universal salvation was reflected in the religious practice of Universalists by their resistance to anything that made claims of difference or distinction among Christians, and hence the lack of any definite religious portrayals in the Goddard Chapel. The indefiniteness of its doctrine also, perhaps, explains the lack of religious vigor that, over time, dissipated the original impulse of the Universalist religion; in 1961 the Universalists combined with the Unitarians to make up the Unitarian Universalist Church, which remains a small denomination but is influential well beyond its numbers. Like most other mainline churches, Universalism has aimed its attention more to social issues than personal salvation; and, among other enterprises, the Universalist Unitarian Church runs the Beacon Press in Boston, which not only publishes books of religious interest but is well known for its titles on social justice and advocacy.

At the time Tufts was founded in the middle of the nineteenth century, just prior to the outbreak of the Civil War, Universalism was a form of vigorous Protestantism that held religious beliefs in contradistinction to the Congregational and Episcopal churches that dominated the cities and town squares, its membership arising from the farmers and small businesspeople of the outlying areas where capital was scarce and life harder than in the more affluent central locations. This isolation made it possible for Universalism to spread more easily from congregation to congregation in the eighteenth century—but not without provoking ferocious opposition from the regnant Protestant churches, which were dominated by a Calvinist theology brought by the Puritans. (A pamphlet of the time in the Tufts University library is entitled, "An account of the putrid Murrinitish plague, lately broke out into the city of Philadelphia, with the causes and symptoms: Also, the manner of treating the infected with an infallible cure.")

Having been founded as a Protestant institution, Tufts had a chapel from the beginning, which, before the erection of Goddard Chapel, was located in the first building erected on the campus, Ballou Hall. (A red brick building, Ballou has stained-glass windows, even though its spire conspicuously lacks a cross.) The lack of a seminary must have bothered the founders and supporters of Tufts, because eventually, in 1869, the Crane Theological School was established; however, it was disbanded in 1968 by the university, having lasted exactly one century. The Tufts website explains: "The closing of the school represented the growing distance of Tufts from its roots in the Universalist religion."[2]

Tufts University Loses Its Religion

During the 1990s, two events occurred at Tufts University that clearly documented its change from a religiously founded and inspired institution to a secular one. The first referred to the fact that Tufts University has maintained, over many years, a Department of Religion, which offers a minor in religion (but not a major). In 1997, however, due to retirements, the number of fulltime faculty in the department had declined from three to one, and despite pressure from interested students, the academic dean (Elizabeth Ammons) refused to commit to the hiring of replacement faculty. The situation was especially dire because the one remaining religion professor, Norman Daniels, had reached retirement age. However, with student and faculty pressure and with the staunch refusal of Daniels to take his retirement even though it was overdue, the administration, it is fair to say, was shamed into hiring two new professors of religion. However, the emphasis is no longer on the Protestant tradition, for the religion courses now being offered beyond the introductory level have a strong multicultural and feminist tendency. As of now, despite its chaplains, chapel, and religion department, it is clear that, by intent, Tufts University has become a secular institution.

The other incident that indicated a downward spiral of religious dis-identification at Tufts clearly revealed that secularization was the preference of its administration. The incident, which made both the on- and off-campus news, came about because the university moved to deny recognition to one of its religious student groups, the Tufts Christian Fellowship. The Tufts Christian Fellowship (TCF) was the only Protestant evangelical religious student group on campus, and the effort to exclude it from Tufts was an indicator of two tendencies: first, the slow erosion of a religious sensibility from higher education (generally) in America; second, how policies aimed at promoting tolerance can themselves curtail religious liberty. Before examining the general ramifications, however, it will be useful to describe the case at Tufts in more detail.

What happened was that one of the TCF's members, a young woman, ran for one of the officers' positions in the TCF; the difficulty with this was that she was a self-announced lesbian who maintained that homosexual behavior is legitimate according to the doctrine of the Christian faith as she understood it. The TCF is a Protestant Christian group affiliated with a larger, nationwide group that provides support, instruction, and leadership for college students who are evangelical Protestants (acting, in this regard, something like Newman Clubs for Catholics or the Hillel House for Jewish students, which also are affiliated with national organizations). To prevent student leadership of the TCF from taking a position on an important moral issue in contradiction to its own essential beliefs, which in evangelical fashion are based on texts from the Bible, the TCF's leaders, who were members of the nationally affiliated evangelical group, forbade her from running. In response, the young woman appealed their action to the Tufts Community Union Judiciary (the TCUJ is a student group) on the basis that the TCF chaplains had denied her rights in direct violation of Tufts University's stated policy that forbids discrimination on the basis of sexual preference. The TCUJ acted promptly and ruled on behalf of the student plaintiff, with the effect that the TCF was now to be denied its status as an official Tufts student organization, which meant that it would lose its office space on campus and its subvention from student-activity fees; also, the TCF could no longer use the Tufts name—all of which meant that the organization was no longer, in effect, a part of Tufts. In turn, the TCF appealed the student judiciary's decision, and the dean of students decided that the judiciary must rehear the case, because due process was not followed the first time: the decision had been made hastily at the end of the spring semester in May 2000. The decision of whether the evangelical organization stayed as an official part of Tufts was, in fact, finally made by the Tufts administration; as of now, the policy of Tufts University disallows discrimination (on the basis of gender identity, race, etc.) in all its various parts, including student groups.[3]

The dispute received extensive treatment in student publications, including *the Tufts Daily*, *the Observer*, and *the Primary Source* (a conservative journal), as well as coverage in the major local newspapers, the *Boston Herald* and the *Boston Globe*.[4] It is noteworthy that the Catholic and Jewish chaplains stated publicly their support for Tufts's nondiscrimination policy, while the Tufts University senior (Protestant) chaplain did not weigh in against the TCF, declaring that its presence on campus was important because it represented a type of religious belief that was not otherwise represented. It is possible that the TCF could have used a different strategy to deal with the candidacy of a declared lesbian who affirmed that homosexuality is not a violation of God's laws; that is, the organization could simply have let her run for the office but then publicly or privately told the other student members of the TCF not to vote for her. The defenders of the young woman intensified the debate on campus with the claim that if the TCF remained a bona fide Tufts student group, the campus would no longer be "safe" for homosexuals.[5]

Interestingly, the above scenario, which took place in 2000, was replayed in 2012. The TCF, which survived the first attempt to remove it from the campus, has once again been "de-recognized" as a bona fide Tufts student organization by the judicial branch of the student government. The ongoing issue, and not only for Tufts University, is at what point does the freedom of religion and association give way to the rights of an individual not to be discriminated against due to sexual orientation?[6]

Observations on the Tufts Example

The first observation is that well before Tufts University disengaged itself from the Universalist version of the Protestant religion, Universalism had itself become hollowed out from within. That is, manifestations of the Universalist Protestant faith on the campus were not systematically sought out and destroyed in some secularist crusade; rather, Universalism itself had become progressively detached from traditional Christian doctrine, for from its origins, as it denied the Calvinist dogmas of predestination and eternal punishment, it also left behind Calvinist teachings about mankind's sinfulness and Christian faith as the means to salvation. The effect was that Universalist worshippers had less need to worship Christ himself or for self-examination of personal conscience, as Christian concern for salvation, over time, became a generalized concern for social justice, reflecting the developing tendency of liberal Protestantism.

The second observation is that the conflict that took place between the Tufts Christian Fellowship and the university's antidiscrimination policies is representative of a much wider conflict, as the same issues now confront the nation at large. Here it is useful to try to state more exactly what that conflict is, given its national resonance. Traditional Christianity has what, in contemporary terminology, would be termed a "rigid" attitude toward sexual morality; the *Catechism of the Catholic Church*, which represents the Church's official teaching, condemns (basically) any sexual activity outside of marriage, and any form of sexual behavior inside or outside of marriage that prohibits the possibility of conception. Protestant attitudes are much more variable, but evangelical Protestantism, while making an exception for birth control (and probably abortions in severe cases), by and large matches the Catholic position, as does Orthodox Judaism.[7] All forms of man-to-man or woman-to-woman sexual acts are strictly condemned as sinful types of behavior by traditional-minded monotheistic faiths. Such an attitude reflects that idea that God has a divine plan or standard that he applies to sexual behavior and to all types of human behavior, and that individual persons are responsible to God for their behavior.

This traditional religious belief regarding sexual morality conflicts in absolute terms with the prevalent ethical notion that non-procreative sexual acts are not only acceptable but should be encouraged as a means of self-expression and self-fulfillment. There seems to be no common point in this conflict that religious believers and advocates of the new sexual morality can accept; is the main purpose of sex (or one of the main purposes, allowing that bonding between a man and a woman is also a discernible purpose) reproduction of the human species? Is sexual activity of *any* sort morally legitimate as long as those engaging in it do so of their own unencumbered free will? To what extent is sexual freedom a right prescribed in law and the US Constitution? Is the recent development of sexual freedom in American society having deleterious effects (e.g., pornography, sexual abuse of children, breakdown of traditional family structure)? Is the new sexual freedom an effective means of developing self-worth for women and men? What is the essence of the Christian religion—is it the tolerance expressed by Christ toward the woman discovered in adultery or the denunciations of St. Paul of wanton sexual behavior? As of now, there are no answers to these questions—or even a means for discussing them—that are acceptable both to defenders of traditional religious morality and defenders of the right of sexual self-expression. The conflict between the secular state and religion in these days is concentrated on the issue of the ethics of sexual behavior (see Chapter 15).

The third observation is that the phenomena of an institution of higher learning founded upon religious principles becoming secular over time is a general one and not isolated to Tufts. Just within a hundred miles of Boston, there are outstanding examples. Harvard University, which began as a Congregational seminary, is now the premier secular university in America. Boston University, which was founded by Methodists in the nineteenth century and which, like Harvard, still maintains a divinity school, is essentially a secular institution with little or no effect of its Methodist origins apparent outside of the divinity school and (possibly) its philosophy department. Brandeis University, despite its Jewish identity, was explicitly secular from its beginning (although when one president of Brandeis tried to remove the Hebrew motto from Brandeis's seal, her effort was successfully resisted; critics pointed out that Yale University has a Hebrew motto on its seal). Brown University in Providence, Rhode Island, which began under Baptist auspices after Roger Williams decamped from Congregational Massachusetts, was headed by Baptist ministers who served as its presidents for the first hundred years of its history; now, however, no sign of Brown's religious origins are to be found on its leafy campus. Dartmouth College in New Hampshire began with the mission of educating Christian missionaries who would evangelize the Native Americans of the region, but again, no actual trace of its religious origins are visible on its campus. The ongoing process of secularization is less noticeable in Catholic institutions but still exists, so much so that Catholic educators have been in conflict with the Vatican.[8] Locally, Boston College, founded by the Jesuit order in the nineteenth century, was upbraided for neglecting its Catholic heritage by Cardinal Law when he was Archbishop of Boston and also by the prominent Franciscan preacher, Fr. Benedict Groeschel.

In all of these examples, defenders of these colleges and universities could argue that the existence of chaplains, divinity schools (BC, BU, and Harvard), departments of religion or religious studies (Tufts, Brandeis), and active student religious groups (including Jewish, Protestant, Catholic, Muslim, and Eastern religions) point to the fact of a healthy presence of religious activity on these campuses. Further, it could be argued that even without explicit religious indoctrination, students are still motivated to higher, idealistic ends by means of incitement to volunteerism and activity on behalf of various social causes. Indeed, Tufts does have a variety of religious groups active on its campus and heavily emphasizes student volunteerism—and not just casually, but formally. Tufts's Leonard Carmichael Society (LCS), named for a past president of Tufts, involves a large number of students in activities such as fundraising for charitable causes, mentoring teenagers, and teaching reading to pupils at the local public schools. Further, the presence of political clubs and student groups aimed at various causes indicate that Tufts students (and college students in general) are not just interested in a vocational training that will enable them to earn a lot of lucre in the current economy.

Religion is available on the Tufts campus today, as its undergraduates have options for religious study and worship across a range of traditions. Ecumenism is encouraged—at one point during the author's time there, Ash Wednesday was celebrated for both Protestants and Catholics, with the priest who served as the Catholic chaplain sharing the imposition of the ashes on the foreheads of penitent Christians with one of the Protestant chaplains. Protestants, Catholics, and Jews have their own university-supported facilities, and there are student clubs for Hindus and Muslims as well. The difference is that formerly, as of perhaps sixty years ago, the Protestant origins of Tufts University would have suffused its culture and constituted its ethos—but now no longer does.

The situation at Tufts, and *many* colleges and universities today, exemplifies the situation as it is evolving nationally and in Western culture. Religious belief has changed from the felt basis of the culture into a personal option, and there is no question that the virtual elimination of religion from the sensibilities of major colleges and universities has been particularly harmful for the appreciation and influence of religion at large. The specified function of higher education is to form the beliefs and opinions of the leaders of American life in economics, politics, information sharing, art, literature, and education; and the elimination of the religious sensibility from higher education has had a trickle-down effect on American culture in general. Despite this process of secularization, religious belief persists in the general culture, for many religious institutions continue to maintain their levels of membership and, in some instances, are experiencing growth in numbers and influence in local cultural environments. The problem is that with the elimination of religion from higher education, there lacks a coherent way for college-educated people to understand the appeal, nature, and persistence of religious belief in American society.

Notes

1. See, for example, http://usatoday30.usatoday.com/religion/2010 (accessed 12-6-2012).
2. The Tufts website is at http://tufts.edu.
3. The Tufts antidiscrimination policy is "... not to discriminate on the basis of race, color, national or ethnic origin, age, religion, disability, gender, sexual orientation, gender identity or expression, or status as a veteran." http://uss.tufts.edu/stuServ/bulletin/Tufts%20Bulletin201; 55.
4. The student newspapers recorded the controversy in detail. See articles and letters in *The Tufts Daily*, April 20, 2000; Sept. 13, 15, 2000; October 10, 12, 13,17, 19, 2000. See also reports in the *Boston Globe*, November 29, 2000, and the *Boston Herald*, April 29, 2000.
5. See *The* (Tufts) *Observer*, Sept. 15, 2000, opinion piece by Adam Carlis, p. 5., "TCF: Not a Safe Space."
6. Katherine Landergan, "Tufts evangelical group faces 'de-recognizing' for requiring leaders to adhere to 'basic biblical truth.' *Boston Globe*, Oct. 27, 2012. http://bostonglobe.com/metro/2012/10/26/tufts-student-group-votes (accessed 11-5-2012). See also Mike Adams, "Things are Tufts All Over." http://townhall.com/columnists/mikeadams/2012/11/5/things_are_tufts_all_over (accessed 11-5-2012).
7. *Catechism of the Catholic Church*, 1994, Libreria Editrice Vaticana, sections 2366–2372.
8. As expressed in a formal document on Catholic universities promulgated by Pope John Paul II in 1990, *Ex Corde Ecclesiae*.

Chapter 4

Three Representative Responses to Decline

Porous Wall of Separation

The wall of separation between church and state in America, so described by Jefferson, has always been porous. The separation doctrine is constitutionally established and has been successful in guaranteeing freedom of religious worship in America without establishing a particular religious denomination as America's official religion. Nonetheless, there has been a fervent devotion to religious belief in American culture from the beginning, and while the connection between church and state has been informal, it has been real in its influence on public, as well as private, morality. Now, however, this informal but real bond between church and state in America is undergoing dissolution; over the past fifty years, there has been an undeniable change in the religious attitudes, beliefs, and practices of the American people. In general, this seems to be due to the decline of the influence of Protestantism in its various forms, which has largely dominated the American ethos. The decline probably began in the 1950s rather than the 1960s, at a time when American self-confidence was at its highest, when social mobility became available to all classes, when self-definition seemed the prerogative of every American citizen and of the nation itself, and when sexual mores were loosening.

The picture of religion's influence in America was never less than complex, as Catholic influences came to the North American continent even earlier than Protestant influences (in the upper Midwest, Southwest, and Southeast); as different Protestant denominations, such as Anglicans, Quakers, Lutherans, or Congregationalists, dominated different regions; and as Jewish congregations were early arrivals as well. But the benign attitude of America toward religion may be best revealed by the number of religions that are, so to speak, native to it: Mormons, Seventh-day Adventists, Southern Baptists, Jehovah's Witnesses, Christian Scientists, and Scientologists. All of these religious denominations originated in the United States and now have adherents worldwide, numbering in the hundreds of thousands—and in some cases, in the millions (see Chapter 2).

Two books and their authors are dealt with in this chapter. Both books deal with the contemporary setting of religion in American life, in the context of the dissolution of the pervasive but informal bonds between Protestant Christianity and America's civic culture. The authors of these books are representative of the extremes of response to the decline of the American religion: Wolfe, the liberal academic, in approval; and Neuhaus, the neoconservative priest, reacting in dismay. Each in his own way describes the characteristics of contemporary religious life in America but proceeds from there to propose an interpretation of the intimate connection between American religious belief and American culture as it exists "after the fall." Finally, although not in the form of a book, Jerry Falwell's response to the decline represents an aggressive attempt to reinstate the *status quo ante*, with a direct attempt to reinsert evangelical Protestantism into the public square. As Sam Tanenhaus has written, "only in America do religious and political ideals become interchangeable, even indistinguishable."[1]

Alan Wolfe Approves the "Transformation" of the American Religion

The Transformation of American Religion: How We Actually Live Our Faith is the second of Alan Wolfe's books about the contemporary situation of religion in American life; the first was *One Nation After All*,[2] which was based on surveys of the attitudes on various issues by religious believers in the United States. Wolfe's thesis in that book is that despite religious differences, Americans have more or less the same general attitudes on such issues as patriotism, racism, welfare, homosexuality, and politics. Wolfe is an avowed liberal, and he took heart from the apparent fact that their beliefs did not so sharply differentiate religious believers from the rest of the general populace—and that believers showed much more tolerance about these issues than what secular-minded citizens might expect. In *The Transformation of American Religion*, Wolfe explores his theme further, making the general assertion that in America, religion itself has been transformed. As he says, "In every aspect of the religious life, American faith has met American culture and American culture has triumphed."[3]

Wolfe is the director of the Boisi Center for Religion and American Public Life at Boston College; and, as a personal note, I will mention that in 2005, I attended a summer seminar run by Wolfe, entitled "Religious Diversity and the Common Good." The seminar was attended by college professors with an interest in religion and was underwritten by the National Endowment for the Humanities. As a teacher, Wolfe was eminently fair minded, not at all resembling the cliché of the liberal professor who will not accept any deviation from his own political convictions. At the beginning of the book, he criticizes the antireligious sentiments of academics "especially in the humanities and social sciences."[4] Wolfe is exceptionally well informed and articulate, characteristics that are represented in *Transformation*. He is a well-known public intellectual who writes frequently for the liberal *New Republic*.

The general method of *Transformation* relies on the techniques of social science that have certain characteristic advantages and disadvantages. (One of the influential pioneers of using social-science techniques to analyze religion in America is Peter Berger, who influenced both Wolfe and Neuhaus.) The great advantage of relying on social science to examine religion is that it allows for a dispassionate comparison of the entire range of religious expression and experience that occurs throughout contemporary United States. Social science attempts to be value free, and so dogmatic assertions peculiar to individual religions are not allowed to dominate the overall analysis. Wolfe relies often on his own observations of worship services and also on interviews, which are interspersed throughout his book, with believers and leaders of churches, temples, and mosques; mainly, he relies on the reports of other social scientists who have studied the field of religion in America, and while there is a very large number of such studies, Wolfe seems well acquainted with the material. Relying on social science to study religion has certain built-in disadvantages as well, however, for a dispassionate approach may leave the researcher unable to understand the power of religious belief to motivate believers, and a value-neutral approach tends to ignore or underestimate the place of doctrine as a means of informing worship and unifying religious bodies.[5]

Transformation is well organized with a constant theme throughout—namely, that despite the rise recently in religious belief, the content of American religious practice is conformable with the beliefs and attitudes of the nonreligious part of American culture; this theme is brought forth forcefully and clearly in the introduction, titled "The Passing of Old-Time Religion," which explains and, in effect, celebrates what Wolfe sees as the decline of the intensity of religious beliefs. The final part of the book, in the conclusion, is titled "Is Democracy Safe From Religion?" Despite the ominous nature of the question, Wolfe concludes that democracy *is* safe from religion, for the subtext in this chapter and throughout the book is what Wolfe thinks is the danger posed to American democracy from the rise of the religious right, personified by Jerry Falwell.

The central eight chapters contain the meat of the book, as each chapter examines a particular aspect of religion; the chapter titles are each one word and readily designate their subjects: worship, fellowship, doctrine, tradition, morality, sin, witness, and identity. Wolfe's procedure in discussing each of the topics is to first define it and then show its transformation from a prescriptive to a more tolerant mode in various religious denominations. Chapter 4, "Tradition," provides an example of Wolfe's method. After a brief introductory section, he tracks the subject from Orthodox Jews to Reform and Liberal Judaism and then to Protestantism and Catholicism. He concludes the chapter with a brief section, "An Innovative God." Wolfe says, regarding the Orthodox, "[I]f Jews emphasize the value of tradition among Americans, the Orthodox emphasize the importance of tradition among Jews."[6] He starts the chapter with consideration of the Jewish faith, because it places such a high value on tradition. He starts with the Orthodox segment of contemporary American Judaism, because it is the segment most concerned with strictly preserving the ways in which the ancient faith has been practiced and understood for millennia. Yet, Wolfe points out, contemporary Orthodox believers are constantly having to interpret their specific traditions in light of American culture regarding dress, segregation of the sexes at worship, kosher food practices, and observing restrictions imposed on Sabbath activities. The degree to which the ancient practices are preserved—and *which* specific practices are preserved—distinguishes Orthodox believers not only from the culture at large but from other Orthodox groups. Wolfe points out that the recent interest in Orthodox Judaism among younger Jews, starting in the 1960s, came about because of its existence as a form of countercultural expression.[7]

Wolfe next treats Conservative and Reform Judaism, but not in detail, using a few anecdotes and quotes to make his point that even as these segments of modern American Judaism retreat from strict observance of ritual traditions, they also attempt to revise the traditions to fit modern lifestyles. Thus, Reform Jews must choose which traditions to ignore, such as sex segregation in worship services, and may argue about the wearing of the yarmulke. Wolfe argues, somewhat against the grain, that even if they reject tradition in general (as in Reform Judaism), that Jews are still dealing with tradition—how much of it to retain in order to stay attached to their identity and heritage, and how much to revise or discard in order to accommodate the pressures of living in today's America. Wolfe's comparative analysis indicates that Conservative Judaism seeks to retain the connection with Jewish religious tradition as expressed in the Torah and in traditional ritual practices, but that Liberal Judaism has discarded the explicitly religious dimension of Jewish identity and replaced it with a zealous concern for activism and social change.[8]

Wolfe proceeds next, in his analysis of tradition, to examine evangelical Protestantism. Here he finds a contradiction at the heart of the recent movement of the revival of evangelicalism and Pentecostalism within contemporary Protestantism. This movement seems to be a return to "old-time" religion, which separates itself from mainline churches by emphasizing the personal experience of salvation; nonetheless, the movement is strongly antitraditional. Wolfe asserts this based on the fact that evangelical churches spring up and are designed to appeal to various currents that arise in contemporary society (e.g., appealing to discontented minority youths in the inner city or to well-established Caucasian business families in the suburbs). Individual evangelical churches are designed to appeal directly to specific social subsets by means of altering worship services and modifying the gospel message itself to fit the needs of that particular set of potential worshippers. "The fact that evangelicals approach their faith in such innovative ways, however, simply underscores how dynamic conservative religion in the United States has become."[9] Wolfe concludes the section on evangelicals in this chapter by pointing out that the recent upsurge in evangelical involvement in (right-wing) politics represents a break from tradition, for had "conservative Protestants adhered to the traditional preference for personal salvation over political activism," we would not have had the mass movement against abortion or, possibly, the presidency of George W. Bush.

Catholics are the last group that Wolfe deals with in the chapter on tradition, and here he shows how the church that is taken to be the paradigmatic example of a religious institution that reveres tradition and has rejected the appeals of modernity has in fact not done so. But

Wolfe's case is easy to make, since the Catholic Church in America has undergone two major shifts in the last two generations. First, the immigrant church of Irish, Italian, and Polish parishes, which continued the old traditions and sometimes the native language of the immigrant populations they served, have, over two or three generations, become integrated into the general American middle-class culture.[10] Second, the major effect of the Second Vatican Council was to substantially change the way in which the Catholic Mass was celebrated as well as to provoke other changes in Catholic practices. Wolfe cites particularly the change in the practice of confession from "the scariest sacrament," which formerly took place in a dark box, to something done more openly and less in the manner of reciting a checklist of the believer's sins than an informal discussion of the believer's spiritual life.[11] Wolfe also cites the less-authoritarian nature of the relation between lay Catholics and the Church's priests and hierarchy; the accommodation of Catholic teaching on marriage and sex to the "realities of modern life"; and the recent influx of Latino immigration, with its emphasis on traditional practices and beliefs, which Wolfe nonetheless confidently assumed will be converted into an Americanized and implicitly less-threatening style of Christian worship.[12]

Wolfe concludes the chapter on tradition with a short summary entitled, "An Innovative God." He rejects the notion expressed in the analysis of political journalist Michael Barone that "religious believers are in the forefront of a determined effort to restore traditional morality to the country."[13] Rather, he maintains that since to conserve is to innovate (reflecting Burke), that while the "United States has conservatives aplenty . . . it lacks traditionalists."[14] Wolfe asserts that the turn of American politics to the right since the time of the Reagan White House does not portend a cultural shift toward traditional morality, but this conclusion may be based in liberal hope. It is said that at bottom, all serious disagreements are theological disagreements about the nature of man and God. Wolfe, however, does not think that insight applies to religious believers in America; differences in ideas about God and man do not engage Wolfe as he diligently does his sociological research on religion in America; nor does he seem to think—or hope—that they engage anyone else.

Richard John Neuhaus Discovers the "Naked Public Square"

Richard John Neuhaus's *The Naked Public Square*, although published over twenty-five years ago, still retains its importance for having given American culture a phrase and an idea that continue to characterize the American religion—that there is a public square in which secularism is dominant, but that, without the influence of religion, that area of civic life is "naked." Neuhaus, who died in 2009, began his career as a Lutheran pastor who was active in the cause of racial justice; he knew Martin Luther King Jr. and ran a pastorate in Harlem, where he looked to the physical as well as spiritual needs of his congregants and the people of Harlem. He was intellectually inclined as well as being a socially active priest; his concern with what he discerned as the collapse of American values, the failure of the "mainline" Protestant churches to effectively confront it, and the rise of "moral majoritanism"—of which he was permanently suspicious—led him to become a serious critic of American religiosity. Eventually, he converted from Lutheranism to Catholicism and was quickly ordained a Catholic priest by Cardinal O'Connor of New York. In his later years, he become, in effect, a religious "neo-con," making frequent television appearances on cable networks as diverse as PBS, C-SPAN, and EWTN to present his view of the necessity of religious ideals in the political and cultural life of America. He was the founder and editor in chief of the religious journal *First Things*, which continues to publish articles of high intellectual quality and for which Neuhaus wrote extensively long editorials.

The Naked Public Square cannot usefully be summarized by describing its chapters or groups of chapters. Neuhaus's style is wordy and ruminative, consisting of a constant flow of references to the movements of the 1980s, such as civil rights and the Moral Majority; the writings of then-current theologians, including John Courtney Murray and Reinhold Niehbur; and political philosophy from Rousseau to Rawls. As Dr. Johnson said of Burke, the stream of Neuhaus's mind "is perpetual." Each chapter contains not a part of the thesis of the book but virtually all of it, since Neuhaus does not construct his argument in parts culminating in a final conclusion, but presents it up front, so to speak, and then examines each of its constituent parts. This makes for a text that is somewhat difficult but nonetheless worth reading, given the range and depth of Neuhaus's knowledge.

The book is best described according to its main theses, of which six can be clearly discerned: secularism is rampant; liberal social attitudes dominate the mainline Protestant churches; the fundamentalist reaction is an unacceptable approach to church-state relations; democratic society is threatened by the recent and continuing loss of public virtue;

the role of culture should be predominant in American society as opposed to law; and a vigorous form of the Judeo-Christian tradition should be imported back into the public square for the public good.

Secularism

The dominance of a diffuse-but-effective secular philosophy in American society, Neuhaus takes as a given, and he is deeply concerned that its proponents are forcing religion out of the public square. By this, he means that explicit references to religion are being systematically forced out of all areas of life that are public (i.e. available to the citizens at large, including the public schools and the courts). He cites evangelical Protestant critics of "secular humanism" such as Tim LaHaye and Francis Schaeffer.[15] The isolation of religion exclusively into the private sphere, Neuhaus sees as unacceptable given the culture-forming and meditative role that religion plays in the social context.[16]

Liberalism of the Mainline

The liberalism of such bodies as the National Council of Churches, as well as the governing elements of major Protestant denominations such as Presbyterian and Episcopal, has been well known for a long time. Neuhaus sees this development in terms of the encroachment of secular philosophy on true Christian doctrine. He is particularly concerned with how the liberalism of the mainline Protestant churches accedes to secularization of the culture, which thus legitimizes the exclusion of religious tradition from the public square.[17] Earlier, the Progressive movement that was concerned with alleviating the condition of working class, African Americans, and the dispossessed could more readily be allied with Christian and Jewish concern for the poor and the marginalized. However, with the social developments of the decades of the sixties and seventies, which legitimated the free exercise of libidinal urges, there has been a departure in the Progressive movement from traditional Christian-Judaic principles of social morality, particularly concerning family life. Now, rather than religion defining public morality, the moral neutrality of the state, with its utilitarian ethics, is defining religious morality.[18]

The Unacceptable Moral Majority

Neuhaus's negative reaction to the rise of the religious right, especially as exemplified in the person of Jerry Falwell, is apparent throughout the book. In fact, *The Naked Public Square* begins in the first chapter with a discussion of the Moral Majority.[19] Neuhaus's reaction seems to be more one of aesthetic repugnance against the vulgarity and bluntness of the late Jerry Falwell than a considered response to the theology of evangelical Protestantism. He considers the technique of the moral-majority movement, which amounts to politicizing a religious cause, to be inimical to democracy, even as its concern about the lack of moral center to American culture and politics are legitimate.[20] The rise of the religious right forces Neuhaus to deal with the issue of the proper relation between church and state, which he believes is very much out of balance.

Democratic Society Is Threatened by the Loss of Public Virtue

In the ninth chapter, "Private Morality, Public Virtue," Neuhaus attacks the secularist attempt to exclude religion from the public sphere directly. He states that that the assumption of the constitutional founders was that religion, actually mainline Protestantism for many years, would underlie the "bare-bones constitutional polity" based on individualism and self-interest.[21] The decline of the influence of the Protestant religion in America was followed by a plethora of negative social consequences that reached a "frenzied apex" in the sixties and seventies, including mass sex murders, an explosion of teenage pregnancies, and abortion.[22] That the manifest evidences of social decay are caused by the decrease of religious influence in American life is plainly evident for Neuhaus.

Culture versus the Law

Throughout the book, but especially in the last chapter, Neuhaus criticizes the reliance on formal law as promulgated by the federal government and in American legislatures and courts. He sees the law being used as a means of excluding religion and public philosophy from the public square and approvingly cites the challenges to laws with this secularist intent by Christian lawyers.[23] For Neuhaus, it is culture that has the mediate role between the state and the individual, and it is therefore that on which Americans should rely for the emplacement of religious and civic values in the fabric of American life. Reliance on the law, Neuhaus implies, can too explicitly define matters in a theoretical manner rather than in a manner based on the history and character of the American people.[24]

Importance of the Judeo-Christian Tradition

The loss of the Protestant ethic in American life that took place in the middle of the twentieth century is the theme that propels *The Naked Public Square*, since in Neuhaus's view, that decline has precipitated the current moral and political crisis in America.[25] Reviving the influence of Protestant ethics is, however, not a plausible approach to revivifying America's moral life; instead, Neuhaus relies on a movement based on a combination of traditional Protestant, Catholic, and Jewish religious influence, encouraged through conversation and serious discourse among the citizens. (To help accomplish this, Neuhaus started the journal *First Things*.) He thinks that the need for a public ethic is self-evident, and the historical fact that American culture was formed—and still *is*—in large if unacknowledged part by the Christian belief system indicates plainly where this public ethic should come from.[26]

The accuracy of the analyses of *The Naked Public Square* can be evaluated from the form of events that have taken place since its publication in 1986. In summary, it appears that Neuhaus's analysis of the dire effects of American culture without a religious basis has become increasingly evident, but that the serious debate about the need for a public ethic that he looked forward to has not taken place. Instead of a movement to restore public virtue, there have come two trends that have obviated the felt need for an extended and serious debate on the issues of the relationship between religion and a public philosophy.

First has been the broad acceptance of the collapse of public moral standards, exemplified in the vulgarity of popular entertainment, high divorce and illegitimacy rates, cohabitation among couples, and the legitimizing of alternate lifestyles (e.g., gay marriage). As each of these developments has been the subject of litigation and legislation, they have become protected by law, and thus it appears that Neuhaus's worry about the dominance of law over culture (and religion) has been plainly justified. More importantly, much of the public no longer thinks of this decline in standards as a crisis to be addressed. Second, the explosion of antisocial trends that Neuhaus reflected on has somewhat abated; divorce rates have stabilized, and general attitudes now reflect a pro-life more than a pro-choice sensibility, so that pro-life politicians no longer fear that their antiabortion stance has doomed their candidacy. The crime rate has significantly declined, even though that has happened, it appears, because the rate of incarceration has also increased over the last twenty years and not because of a rise in the appreciation of the need for public virtue. As a result of these trends, the debate about the need for a religious revival and the nature of a public philosophy that would tie together the individual and the public order is found only in academic circles and no longer in the middle-brow journals, in news accounts, or online.

A noxious effect of the secularization of the public square has been the substitution of political for religious categories of public morality, such that belonging to a religious body that accepts the traditional or biblical standards of morality has become a political act. Believers in traditional religious morality are termed "conservative" or "right wing" by the secular culture; correspondingly, eschewing organized religion and accepting the freedom now protected by expanded legal standards of behavior—such as the elimination of sodomy laws—identifies one as politically "liberal" or "progressive." The conflict between conservative and liberal belief systems continues to dominate mainline religious organizations; a regrettable example (as examined in Chapter 2) is the current schismatic conflict taking place within the Episcopal Church.

Falwell *Agonistes*

The two books considered here testify to the extremes of religious thought in America, post-decline, but although they express two very different viewpoints, there is a major similarity. Both authors were energized by the appearance of the religious right that took place in the 1980s; perhaps surprisingly, Neuhaus was initially as opposed to it as Wolfe. They were faced with a direct attempt to use religious belief as a force in partisan politics—the "Moral Majority" movement exemplified in the person of the late Baptist minister Jerry Falwell. The repugnance that Falwell incited in his critics was notable, for he was effective enough to be feared by liberals who were pressing their agenda on the national scene, but it was his persona that seem to mainly put people off, including some of those who agreed with him in essence. Falwell came on as a performer, loud and unapologetic; his appearance, manner, and speech patterns were not those of a person educated in the northeastern United States; in fact, he was a southerner who had been a "bad boy" in his youth, had undergone the typical Baptist experience of conversion to Christ, and had received his education not at an elite university such as Duke or the University of North Carolina, but in bible colleges.

Falwell was confrontational and did not explain his position in measured tones, which made him, however, an effective spokesman for traditional Protestantism on television and in many live venues, including a contentious appearance at Harvard University. Falwell's saving grace as a public figure was his generous sense of humor; when Senator Edward Kennedy, a prominent and influential liberal, gleefully announced that he had (accidentally) received one of Falwell's fundraising letters, Falwell responded by inviting Kennedy to appear with him in front of a student audience at Liberty University. Both men, notable "hams" (i.e., experienced public performers), teased and joshed one another in their appearance together; the event was not confrontational.[27]

Falwell was the pastor of a large church in Kentucky and founded his own institution of higher education, Liberty University (still a going concern), but mainly he was known for being the founder of the Moral Majority movement that attempted to transform the traditional Protestant religious sensibility into a political movement, which was seen by its critics as an unwarranted and dangerous transgression of the separation of church and state. From Falwell's point of view, religiously inspired political action was justified as a response to the forces of secularization that were pursuing an explicitly antireligious agenda in American culture, law courts, and legislative bodies. Falwell eventually left off his activity in the Moral Majority movement to move back to preaching, pastoring his church, and running Liberty University. Unlike Pat Robertson, who was also a prominent figure in the religious and political right, Falwell never attempted a takeover of the Republican Party.[28]

Falwell's career as a public figure was a harbinger of the direct intrusion of traditional Christianity into American politics. The Moral Majority group was later led by Ralph Reed but has since dissolved; but the thrust of Falwell's Moral Majority movement has, if anything, become more prominent and influential. Charles Colson and other evangelical Protestants later engaged traditional-minded Catholics in their movement, including, eventually, Neuhaus himself; his journal, *First Things*, includes features from Protestants, Catholics, and Jews who view the secularized political culture of America with dismay.[29] Meanwhile, evangelical Protestants and Catholics have become prominent supporters of some Republican politicians, including George W. Bush and, more recently, Rick Santorum. Indeed, the support of the evangelical voting bloc has become essential for success in the Republican Party on the national scene.

Neuhaus is the gloomy prophet of the three responders (perhaps less optimistic than Falwell that religion can reenter the public square), perceiving the dissolution of church from state from the point of view of traditional Christianity; but he seems to have been the most accurate in predicting its effects. Because of its exclusion from the public square, traditional religious activity has been moving inward rather than attempting to influence the culture in general, as if recognizing that religion has been so thoroughly excluded from the public square that American culture can no longer be directly influenced by the doctrines of traditional religious bodies. Within traditional and orthodox religious groups, there has been a sharpening of appreciation of their traditional discipline, doctrine, and practice as if the effort at culture forming is wasted, so that at this point in American history, it is better to cultivate one's own religious garden.[30]

In contrast to both Neuhaus and Falwell, for Wolfe, the decline of American religion and the process of secularization is a good thing. Wolfe concludes from his sociological research that there has been a lessening of the intensity of religious belief in America, a fact that he finds reassuring, since religious believers of a traditional bent are more likely to be intolerant of the secularization of American culture. Wolfe's liberalism is consistent with Enlightenment ideals of toleration, which leads him to regard religious belief as an irrational and potentially uncontrollable force inimical to democratic politics. Wolfe does not deal with the correlation between the decline of the informal religious-belief system in America and the subsequent decline in public moral standards highlighted by Neuhaus. Neither does he ever seem to consider the possibility that religious belief is directed to an objective spiritual reality, in obvious contrast to Falwell. For Wolfe, scientific rationalism is inherent in his choice of methodology for studying religion in America, and he seems also to have imbibed the antireligious spirit that has been a part of the scientific mindset since the time of Galileo.

The ongoing process of exclusion of religion from the public square has helped foment a cultural and political divide in America that has too easily slid into the contemporary context of "red state versus blue state" politics. For the sad situation nowadays in America is that to be a believer in any form of traditional monotheistic religion—Judaism,

Christianity, or Islam—is automatically to either be or *assumed* to be a "conservative." On the other hand, to be a nonbeliever also slips immediately into political categorization; to be a nonbeliever or even somewhat agnostic implies a liberal political identity.

In the early days of America's participation in the Second World War, a Catholic naval aviator, Commander Shea, sent a letter to his son (Shea later died in the war), in which he wrote, "Be a good Catholic and you cannot help but be a good American." Sixty years later, however—ironically—with the decline of Protestant morality as the basis of American culture, and with the dominance of secularism, this may be no longer true.[31] If by a "good" Catholic we understand someone who subscribes to the fundamental tenets of Catholic Christianity (which does not exclude liberals, conservatives, or those of no political adherence), then Commander Shea's confidence is sadly no longer justified. And what is true of Catholics is true for traditionalist believers of all religions. For today, whatever their political views, a "good believer" (i.e., a faithful Catholic, an evangelical Protestant, or a Conservative or Orthodox Jew) will not be seen as a "good American" by an increasingly influential segment of American culture and politics.

Notes

1. Sam Tanenhaus, *Whittaker Chambers*; New York, Random House, 1997; 468.
2. Alan Wolfe, *One Nation After All*; New York, Penguin, 1999.
3. Alan Wolfe, *The Transformation of American Religion*; New York, Simon and Schuster, 2003; 3.
4. Ibid., viii.
5. See ibid., 67–95 for Wolfe's analysis of religious doctrine.
6. Ibid., 99.
7. Ibid., 103.
8. Ibid., 110.
9. Ibid., 117.
10. Ibid., 118.
11. Ibid., 120–121.
12. Ibid., 119–125.
13. Ibid., 125.
14. Ibid., 126.
15. Richard John Neuhaus, *The Naked Public Square*; Grand Rapids, MI, Eerdmans, 1984, 199.
16. Ibid., 21, 141.
17. Ibid., 235–237.
18. Ibid., 141.
19. Ibid., 6.
20. Ibid., 107.

21. Ibid., 140–141.
22. Ibid., 33, on abortion.
23. Ibid., 248.
24. Ibid., 249.
25. Ibid., 237.
26. Ibid., 116.
27. For Kennedy's speech, see http://www.americanrhetoric.com/speeches/tedkennedy. For Falwell and Kennedy's friendship, see the article in the *Los Angeles Times*, http://articles.latimes.com/1985-02-06/news/mn-43c (both accessed 12-5-2012).
28. Michael Sean Winters, *God's Right Hand: How Jerry Falwell Made God a Republican and Baptized the American Right*; HarperOne, 2012.
29. See Colsen obituary, *Boston Sunday Globe*, April 22, 2012.
30. See "Calvinism is Back," *Christian Science Monitor*, March 27, 2010. www.csmonitor.com/USA/Society/2010/0327/Christian-faith-Calvinism-is-back.
31. Shea was a graduate of Boston College for whom an athletic field was formerly named on that campus. It is today covered over by buildings.

Part II

The State of Science and Culture Now

Chapter 5

The Actual Origins of Modern Science

Whig History of Science

There has been a longstanding controversy about the origin of modern empirical science. Science, as we understand and recognize it today, began in the late seventeenth and early eighteenth centuries and has since progressed to attain its full flower, not only by the accretion of newly discovered facts and the development of more comprehensive theories but in its social effects. On the intellectual level, scientific theories have become a touchstone by which all and any phenomena must be apprehended and explained; for example, human behavior is not explained in moral terms but as the result of genes, society, or early upbringing. Technologically, applied science has radically changed the way we live our lives; one hundred years ago, most Americans did not have indoor illumination, much less a means of instantaneous communication or an independent means of transportation. As a direct result, we have, over the last several centuries, been experiencing the replacement of religion by science as the chief culture former of the West. If this is so, and evidently it is, then it is of importance to understand the nature of modern science more fully than is usual these days, and often the best way to understand the nature of a thing is to understand its origins. So how did modern science itself originate—that is, what were the proximate causes?

There is an ideological account of the increasing dominance of modern science that is often accepted—what can be termed the standard or "Whig" view (the term "Whig history" was coined by British historian Herbert Butterfield to apply to those historians who, like Macauley, assumed that a general progress that tends toward a contemporary situation of individual freedom and political liberty dominates the history of mankind[1]). The standard or Whig view is that the Middle Ages replaced the Dark Ages when the Church had suppressed human freedom and despotically ruled the culture by imposing theological strictures upon those forward-looking thinkers who would rather look to the things of the living earth than to the mythological heaven above, who in this manner replaced barren faith with fruitful results. In this view, the rationalism of the Greek philosophers and mathematicians showed the way to human progress, both intellectually and technologically, but the movement toward rationalism and human freedom was suppressed and sidetracked by the theological doctrines of the Catholic Church, which imposed its doctrines in a theocratic manner and which thereby foreclosed human progress for a millennia. The medieval theologians, it was said, produced crabbed works detailing irrelevant controversies, such as how many angels could dance on the head of a pin. In the meantime, the Catholic Church was, as at the time of the Inquisition, burning heretics, such as Giordano Bruno, in order to keep newly advancing thought and human freedom at bay. But in the end, the efforts of the dark dominion failed, and heroes of thought, such as Galileo, arose, who set the path of modernity going.

Recent scholarship, since the 1970s, has largely disqualified this account, which is really a kind of myth; Galileo, for instance, was not the first scholar to investigate motion and to apply mathematical principles to explain it, for he had predecessors, priest-scholars who had diagrammatically distinguished between motion that was accelerated and motion that was constant, implying a doctrine central to modern science, that of *inertia*. (The distinction between motion that takes place at a steady rate and motion that accelerates is required to discern when external force has been applied to a moving body.) This analysis of kinds of motion was going on the fourteenth century, even as other priest-scholars were arguing about the Immaculate Conception of the Virgin Mary. Muslim scholars, as early as the eleventh century, were analyzing Aristotle's theory of motion and devising important experiments in optics, learning that eventually made its way into the universities of Europe. In the Whig view, the beginning of modern science was, in effect, a revolution that overthrew the constraints of medieval social organization and Christian doctrine imposed by the Church. Stephen Jay Gould, in one of his last books, pointed out the sources of this view of the origins of modern science in the late nineteenth century, including a book written by the first president of Cornell University.[2]

Thus, we may ask on a less general level, what were the actual conditions on the ground, so to speak, that impelled or permitted the rise of modern science in the sixteenth and seventeenth centuries? It is possible, by uncovering these on-the-ground conditions, to explain the origins of modern science in less intellectualized and world-historical terms and to make a point about its origins that is not usually considered. Independently of church doctrines, new patterns of thought, and social changes, the fact is that modern science could not have evolved until certain, more practical conditions had evolved—in materials science and mathematics.

Material Conditions

The development of materials technology was an essential precondition of the evolution of modern science. For example, the measurement of temperature and air pressure requires the ability to make glass tubes to the length of several inches or feet. But this required, in turn, the development of glass-blowing techniques that were unavailable to the medieval technicians who blew the glass for the stained windows of Chartres Cathedral, requiring, besides glass blowing, a very flat metal surface on which to roll a jar of molten glass into tubes. The ability to make elongated glass tubes was required in order for Fahrenheit in the early eighteenth century to make the first usable thermometer and to establish an accurate scale of temperature. Even in the most basic operations of measurement, the development of hard metals—of iron and steel to replace copper and bronze—was a necessary development to define temperature quantitatively and that laid the basis for eventually developing theories of heat in terms of atomic activity.

What Greek empirical science lacked, even as Aristotle and Theophrastus did excellent descriptive biology investigating animal and plant life respectively, was the ability to precisely measure and observe what they were studying. Simply speaking, they had no microscopes by which to examine the inner structure of living things; the existence of cells was only discoverable after Van Leevenheouk, in the sixteenth century, used a piece of ground convex glass to look at specimens with. Up to that point, biology could only deal with surfaces, and the best that could be accomplished on the theoretical level were intense schemes of classification, identifying by comparison the relationships among separate species of plants and animals, as Linnaeus did in the eighteenth century.

The ancients very carefully tracked the paths of the stars and planets, the moon and the sun, as they traversed the sky in repeated patterns, but it was only after Galileo used the newly invented telescope for astronomical observation that the existence of Jupiter's moons and Saturn's rings became discoverable, not to mention the craters of the moon and the rotation of the sun. The revolutionary schema of Copernicus (in two senses of the word) was possible because of the increasing accuracy of astronomical observations over time; large astrolabes made of metal, less subject to change and decay, replaced smaller wooden ones, so that it became possible to measure an arc with more precision, to distinguish degrees and then break them down into minutes and seconds of arc.

The measurement of time was especially important as the basis for the new science of mechanics, which was based on the distinction between the inertial and accelerated motion of bodies. The history of time measurement extends from ancient sundials and hourglasses filled with fine sand into medieval times when the monks invented various devices such as water clocks, since they wanted to rise early to say morning prayers (*matins*) before the sun came up. When he first studied the motion of pendulums, Galileo was forced to use his own pulse to measure their periods; Newton, as a boy, made a mechanical clock by carving the wheels out of wood, a standard project for intelligent boys in his day (just a boy today might design his own video game or make his own computer). Huygens, Galileo's contemporary, invented the first pendulum clock, and Hooke a short time later invented the balance spring for watches. It was only with the advent of mechanical timepieces that accelerated motion could be quantitatively defined, so that Newton's famous definition of force as related to mass and acceleration could be applied to real-life situations and become of practical consequence. It is fair to say that it was only with the development of the material basis for refined measurement and observation, available only after centuries of development and discovery, that modern science could emerge.

Mathematical Developments

Of equal importance as the development of materials technology, without which the emergence of modern science was impossible, were developments in mathematics. Archimedes, the most famous ancient engineer, used geometrical diagrams to explain his various discoveries and laws, basing quantitative description on the areas of squares and triangles. Galileo, who, not incidentally, was employed as a professor of mathematics, also used the same kind of geometrical explanation, so it appears as if there had been no significant development in mathematics as the basis for expressing physical laws from ancient times up to Galileo's time. Today, when Galileo's law of falling bodies is explained in high-school physics classes, it is by means of an algebraic formula expressed as an equality between the distance an object falls and the amount of time that it has fallen. Actually this is incorrect, for the student must learn that it is the *square* of the time that relates directly to the distance, not the amount of time in the first level of direct measurement. Expressing the square of the time can be done by geometric means by literally drawing a square; it is obviously more compact and succinct to express the relationship by means of a single expression using a superscript, as in X^2. Algebra depends on using letters instead of numbers, because they can accommodate any number necessary in a particular application. But algebra was not invented until the tenth or eleventh century, by Arab scholars, a discovery that came eventually to European centers of learning but of which Galileo either did not know or chose not to avail himself.

Other developments in mathematics during the early modern era included the study of trigonometry, which was necessary for studying the motions of stars and planets in a precise enough way to be able to predict their motions and make navigational charts. Logarithms were developed by Napier as a means for quick calculations of large numbers, useful in commercial purposes as well as scientific studies.

The next large step in mathematics came when geometry (founded by the Greeks) was combined with algebra (founded by the Arabs) in the invention of analytic geometry by Descartes. Modern high schoolers discover this in the ninth grade, when the instructor either laboriously draws a diagram on the blackboard or, more often, uses a computer projection to demonstrate the uses of the Cartesian coordinate system, with a vertical "Y" axis and a horizontal "X" axis. Point positions can be identified on this chart by giving numbers along the two axes, but the harried student no sooner learns this than she next is instructed how to draw first straight lines and then curves on the Cartesian coordinate system that are exact expressions of algebraic formulas. (Nowadays these diagrams appear on the students' handheld computers.) The point of the exercise is often lost on the high-school students, however, for by this combination of algebra and geometry ("analytic geometry"), scientific formulas can be graphically pictured in precisely quantitative terms. Relationships between measurable physical entities, including, most significantly, cause-and-effect relationships, can be explained—and, in fact, can be used to predict the future course of physical processes.

The next great advance was inevitable once the curves that expressed, for example, the parabolic path of a cannon ball, could be described by a formula that described the path. For by means of the differential calculus, at any point on the mathematically defined curve, the exact speed of the cannon ball, or its exact angle of upward or downward movement, could be determined. The inevitability of this development of the differential calculus is illustrated by the fact that Newton and Leibniz discovered the calculus at the same time (which resulted in an ugly priority debate). The calculus is the most typical and useful mathematical tool of science, since it allows for precise determination of the status of any physical process that can be measured and quantified at any one time in as many dimensions as the scientific field may require. There is an implied philosophical aspect to use of the calculus which seems to assert that given precise knowledge of the beginning of a physical process, the later and final stages will necessarily occur as predicted; in short, the use of the calculus can reinforce the impression that physical processes are determined in their outcome, and that, theoretically speaking, it is possible for science to explain everything that happens in the universe (this was Laplace's conceit). And with such an impression comes the hubris that is the inherent temptation and typical moral failure of scientists, for the graphical representation of physical processes, with their exact formulas and precise curves, can give a false impression that scientific knowledge completely explains the process even when it does not; thus it is useful when describing the parabolic arc of a cannon ball to know who is firing at whom, and why.

Discarnation

The term "discarnation" that is described in this section somewhat resembles Taylor's use of the term "excarnation."[3] Taylor describes a historical process by which religious elements of understanding the physical universe are gradually removed by the incursions of modern science. In this account, however, discarnation describes a more rapid process that occurred at the time of the arrival of modern science, that is, in the seventeenth century. As used here, the term has not only a religious implication but scientific and philosophical ones as well. Discarnation refers, as background or subtext, to the incarnation of Christ who, as both God and man, brought the ideal and spiritual portion of universal existence down to the physical and material level. This is the primary fact or doctrine of the Christian religion and reflects the Judaic concept of divinity as well, since in the Hebrew Bible, the Lord is an active—not to say domineering—participant in human affairs, unlike Aristotle's concept of God as "unmoved mover," which is a primarily philosophical and not religious idea. There is a long and continuous history of Christian meditation on the incarnation, which is frankly called a mystery; however the participation of the divine in the human is almost always presented in moral, not metaphysical, terms. In presenting the issue in moral terms, the Christian understanding of reality is that the encoding of the ideal and spiritual realm within the physical and material is the result of a choice made by the triune God. Incarnation in *this* sense implies an intimacy between the divine realm and the material world, as if to say that if God as Christ could take on the form of mortal man—that is, if divinity could conjoin with the materiality of human flesh—then the physical universe, of which human nature is a part, is not alien to the presence of divinity.

The overall Christian concept of the relation of the divine to material creation expressed in the doctrine of the Incarnation can be compared to the philosophical account of the relation of general ideas to material instances expressed in classical metaphysics; more relevantly, it can be compared to the general relation of scientific theories and experimental observations. Here we perceive, in the operations of modern science, that the intimate relationship between the divine and the material expressed in the Christian doctrine of the Incarnation has been dismissed. As a consequence, the relationship between theory and experiment in science has become problematic, for it is now common to speak of "models" rather than theories in scientific investigation, as if to imply that no final understanding of the cause of physical events can ever be known. This description is not merely a matter of analysis or speculation, since it also reflects directly the historical process from the medieval worldview to the modern, scientific worldview. But what is seen in the Whig history of science as a form of intellectual and moral progress is, in fact, a process of leaving behind the intimacy of mind and matter implied by the central doctrine of Christianity.

It may seem as though in inventing the term "discarnation," the author is stretching the point, but its usefulness can be seen in the evolution of scientific philosophy, from Bacon to Descartes. The Baconian philosophy of science emphasized the empirical side of science, describing science as a process of discovery working from observations and experiments upward to low-level laws and generalizations. Bacon says nothing about theories as such, except that they ought not to be asserted beforehand in the process of scientific research. It was in this sense that he dismissed Aristotle and the grand constructs of medieval philosophy as theorizing from first principles that had never been properly tested. The inductivist ideal asserted by Bacon did not imply, however, that the phenomena observed came about because of the operations of an immaterial law. The next development in the philosophy of science, a century later, was the Cartesian discovery of the essential importance of mathematically defined laws for the development of modern science. But the metaphysical result was the lack of fit, or of "intimacy" between the theoretical and the material realm as dealt with in modern science.

In denying the options of Platonic and Aristotelian philosophy, which have descriptions of material objects as either a reflection of an ideal pattern or as a combination of matter and immaterial form, and rejecting as a matter of principle the Christian sense of incarnation, the modern scientific sensibility is left with an oppositional sense of the relationship between theory and experiment. The net result of this process of scientific discarnation, of the separation of the divine from the material, is famously Cartesian dualism that results in an irresolvable displacement of the theoretical from the material. *Mind,* in Descartes's philosophy, is derived separately and completely apart from the knowledge of material reality; in Descartes's *Meditations,* he describes an intellectual process by which once having secured the reality of mind, we then proceed to the certain knowledge of the existence of God. Our knowledge of material objects thus appears secondary from our essential knowledge of self and God, for it is God whose power and solicitude certifies that our knowledge of the material world via the senses can lead to correct understanding of the reality of the physical universe.[4] This dualistic metaphysics became the basis of understanding modern science, because it asserted that the scientific discovery in effect is a guaranteed source of certain knowledge of the physical universe. It is this assertion that science is a source of certain knowledge of an independently existent material world that was criticized and rejected by Rorty in his influential book

Philosophy and the Mirror of Nature, which is basically an argument for epistemological relativism.[5]

The History of Scientific Progress, in Three Stages

Understanding the history of science is essential to understanding the nature of modern science. Looking at modern science as a whole, it is useful to divide its explanatory territory into three large areas: physical science, including astronomy, physics, and chemistry; biological science, including genetics, anatomy, physiology, and evolution; and social science, including psychology, sociology, and economics. (These categories are not completely inclusive.)

What is of interest is that this division can be seen in two aspects, for the tripartite division can be understood either in ontological terms of levels of reality or in historical terms of the chronological sequence in which each large field has developed. Historically, it is readily apparent that astronomy and subsequently mechanics (the study of bodies in motion) were the first of the modern sciences to develop in the sixteenth and seventeenth centuries. The efforts of Galileo, Kepler, and Descartes, among many others, culminated in Newton's comprehensive, the *Principia Mathematica Philosophiae Naturalis*. Chemistry slowly evolved out of alchemy—the two were at first barely distinguishable in the sixteenth and seventeenth centuries—roughly culminating in the work of Lavoisier in the late eighteenth century. Biology gained its scientific status, proceeding from a teleological to a mechanical mode, with Darwin's theory of evolution in the middle of the nineteenth century, which was also a time of great activity by professional and amateur naturalists in discovering and classifying previously unknown species, including extinct ones. By the end of the late nineteenth century and into the early twentieth, modern social science, including psychology and sociology, were, in a manner, invented by such figures as William James, Sigmund Freud, and Emile Durkheim.

The historical development from physical science to biological science to social science was not accidental. It is apparent that the stages of development of the three major branches of science reflect the nature of their related subject matter, which involves their *ontology*. Chomsky makes a very useful comment:

> If you take a look at the progress of science, the sciences are a kind of continuum, but they're broken up into fields. The greatest progress is in the sciences that study the simplest systems. So take,

> say, physics—greatest progress there. But the physicists have an advantage . . . if something gets too complicated they hand it to someone else. If a molecule is too big, you give it to the chemists. . . . if the system gets too big, [the chemists] give it to the biologists. And if it gets too big for them, they give it to the psychologists, and finally it ends up in the hands of the literary critic.[6]

Ontologically, the physical sciences deal with the most common elements of physical reality, namely matter itself in all its permutations and basic constituents, for matter underlies all physical reality and is the basis of our understanding and perception of the universe. We depend on our senses, which are, in essence, the material conduits (via light, sound, smell, etc.) to take in the parts of, and participate in, the physical universe. Oddly, perhaps because of its omnipresence, matter is very difficult to describe and categorize in detail, for only partial views of it seem available to us. Aristotle described matter as one part of the individual substances we experience and see around us, since in his philosophy, matter was combined with immaterial form in such a manner as to give individual substances their shape and other qualities; form is also, in Aristotle's philosophy, the immaterial element by which the human mind, equally immaterial, intellectually comprehends individual substances. Not matter itself but rather its *behavior*, as the result of forces acting upon it, was the subject of Newton's mechanical system, which was largely mathematical; as for the inner nature of matter, that is the subject of later discoveries by atomic science, which reaches into the insides of atomic structure by means of quantum mechanics.

The next level up, ontologically speaking, from basic matter is that of living things, including plants and animals. Here, science deals with that portion of material reality that shows "signs of life," of self-growth and activity, for plants grow on their own in the right conditions while fish, amphibians, and land animals also move on their own, performing a variety of life-sustaining functions, including eating and reproduction. The complexity of living things adds another layer, so to speak, to the thickness of physical reality, and while the matter involved in the makeup of living things is only a very small part of all the matter on Earth—much less our galaxy or the entire universe—living matter is more interesting because it is more relevant to human existence. We may be made up of atomic particles at the most basic material level, but it is the biochemistry, anatomical structure, cellular biology, and neurology that defines what our bodies are as living human beings. Ancient studies of living beings are descriptive, but in the nineteenth and twentieth centuries, it became possible to measure and quantify biological phenomena beyond the level of the unaided human eye, and so it is at this point that biology became a modern science, just as the application of telescopic observation and modern mathematics turned astronomy into modern, scientific astronomy. However, it also noticeable that biological studies provoke controversy in a way that physics and chemistry do not, for it matters a great deal if human nature is solely explainable by its biological basis and thus not particularly different from the animal life forms to which the human body is obviously related.

The final level up—ontologically, the final layer of material reality that modern science studies—is the conscious behavior of men and animals. Politics and human behavior have been the study of the human race as far back, at least, as written records exist. The *Iliad* or the *Book of Gilgamesh*, Genesis, and Job, along with the plays of the Greek playwrights—not to mention the well-developed political theories of Plato, Aristotle, the Stoics, and Cicero—all attest to the extreme interest that human behavior has had for human beings from the beginning of human cognition. However, in the process from early accounts to those of modern political science, we do not notice a necessary improvement. Human behavior starts with its basis in human biology, but in proceeding from the biological to the social level, a whole new level of complexity comes about. While there is no doubt that the accuracy of modern biology exceeds by far that of Aristotle's anatomical investigations because of our knowledge, for example, of cell biology, there is significant doubt whether the studies in the *Journal of the American Political Science Association* explain the actions of governments or the motives of political men and women any better than Aristotle's *Politics* or Plato's *Republic*. This fact is illustrated by the existence of a kind of countercultural stream of scholarly study in political science inspired by the controversial figure of Leo Strauss. Strauss, in contrast to the statistical approach used by political science, utilized instead the study of the texts of political philosophers including Rousseau, Maimonides, and Hobbes.

As each of the three broad levels of scientific study is considered, two implications become apparent: first is the discernable pattern of increasing complexity, and second is the pattern of increasing resistance and of resultant controversy. While no one is offended these days by the replacement of astrology and ancient myth by modern astronomy, the deterministic implications of genetic biology are subject to constant critical review, as are the materialistic implications of evolutionary biology. At the top level of complexity, results of the research in the social sciences are frequently contradicted not only by humanistic professors or religious authorities but by the research results of other social scientists. Plainly, the cultural, political, and religious preferences of social-science researchers often influence their particular subject of study, the manner in which they approach them, and the types of evidence they survey. The results of social-scientific studies are routinely attacked within and without the community of social-science researchers, so much so that confrontations between competing schools is reflected in contemporary politics—for example, when free-market economists are cited in opposition to Keynesian economists and vice versa.

Notes

1. Herbert Butterfield, *The Whig Interpretation of History*; New York, Norton, 1965. See also Butterfield's *The Origins of Modern Science*; New York, Free press, 1957; Chapter 12, "Ideas of Progress and Ideas of Evolution," 223–246. It is possible that Butterfield's reflections on the history of science led him to the Whig history thesis.
2. Stephen Jay Gould, *Rocks of Ages*; New York, Ballantine, 1999; 99–104.
3. Richard Taylor, *A Secular Age*; Cambridge, Harvard U. Press, 2007; 613–615.
4. Rene Descartes, *Meditations*; the argument about God and sense knowledge is in the "Fourth Meditation."
5. Richard Rorty, *Philosophy and the Mirror of Nature*; Princeton, Princeton U. Press, 1979.
6. Yarden Katz, "Noam Chomsky on Where Artificial Intelligence Went Wrong," *The Atlantic* (November 2012); http://theatlantic.com/technology/print/2012/11/noam-chomsky-on-artificial-i.

Chapter 6

Athens and Jerusalem in the Twenty-First Century

Athens and Jerusalem in the History of Theology

In the third century, the dissident Christian theologian Tertullian asked, rhetorically, in the midst of a theological controversy, "What has Athens to do with Jerusalem?"—condemning, in effect, the use of Platonic philosophy to defend the Christian religion and provide an intellectual basis for its theology.[1] The theme of the two competing cities has characterized the relationship of the Christian religion with Western civilization even to this day, and according to Leo Strauss, Western civilization attains its vitality and uniqueness because in it two major sources of knowledge and inspiration contend, namely, the secular and the revealed.[2] Ironically, it has been the fact that integrating revealed knowledge—as found in the Bible and religious tradition—with secular knowledge has never been actually accomplished that provides the essential motive force for the advance of Western civilization, at least so far. In the third century, the form that secular knowledge took was the Neoplatonic philosophy that Hellenistic culture of the time inherited from the Greeks, hence Tertullian's reference to "Athens." This was the same Neoplatonism later integrated into Christian theology by the great Saint Augustine. However, while the source of revelation remained constant (i.e., the Gospels and the authority of the Western and Eastern bishops of the Christian Church), the source of secular knowledge changed, from ambient neo-Platonism in the late ancient world to rediscovered Aristotelianism in the high Middle Ages.[3] In modern times, however, secular knowledge has been represented not by ancient philosophy, but by modern empirical science, and the Athens-Jerusalem conflict has continued under the rubric of "religion versus science."

Christian theology has, throughout its history, starting at least in the third century, attempted to explain Christian doctrine in terms of current philosophical thought. The same has been true for the other monotheistic religions as well, particularly Judaism and, to a lesser degree, Islam. In this enterprise of assimilation and explanation, theologians and other commentators have been somewhat successful; it could be argued that the effort began not in the third century but in the very beginning of the Christian religion. Tertullian, in effect, loses the argument about Athens and Jerusalem because they have been intertwined from the beginning of the Christian religion, from the time when the fourth Gospel was written (probably in the early 90s, CE). Thus the famous introduction to the fourth Christian gospel, "In the beginning was the word, and the word was with God and the word was God, and through Him all things were created" (John 1:1)—the term "word" in English translates the term "logos" in the ancient Greek, which has resonances that are deeply philosophical and go beyond the syntactical. *Logos* involves knowledge and reason and ultimately the idea that reality is explained by—because *formed by*—reason, which is understood to permeate all levels of existence. To refer to Jesus of Nazareth as the *Logos* was thus a means of introducing the premier object of Christian worship to a nonbelieving world in a manner by which his singular importance could be indicated.

A review of attempts by Christians and other monotheistics to present their religion in philosophical terms would include a number of serious thinkers, but here we will concentrate on the two best known, Augustine and Thomas Aquinas. Augustine, living in Rome and Romanized North Africa during the time of the empire's collapse into senescence, took it upon himself to understand the Bible and Christian doctrine and to explain their apparent contradictions and intellectual difficulties. He found in the metaphysical vision of Neoplatonism the idea that reality existed not only in material form but in immaterial and spiritual form. This enabled him to conceive the nature of God in ways that went beyond the physical categories of time-boundedness and matter, and yet retain intellectual coherence. That is, God could be infinite without having to be the largest physical thing or the most physically powerful force in existence, or be eternal in the sense of pre-existing in the time before the universe was created. Platonic metaphysics also enabled him to imagine nonexistence as well; thus, Augustine solved the issue of what God was doing before He created the universe, by asserting, fifteen hundred years before Einstein, that prior to creation, time did not exist, implying that time and matter were coexistent things. This insight enabled Augustine to parse the first verses of Genesis to explain creation occurring *ex nihilo*,

The utilization of Greek philosophy continued until nearly a millennium later, when European civilization had undergone a sea change from the remains of the classical Greco-Roman to the high medieval. Thomas Aquinas lived during the height of the Middle Ages, producing a vast collection of what, properly speaking, are theological treatises, but which nonetheless utilized a highly developed notion of Aristotelian philosophy, plus an ability to apply logic to theological arguments that may be unsurpassed in Christian thought. The use of Aristotelian texts, which had reached Europe originally by means of Arab translations, was critical for the development of Christian medieval thought, each major theologian at the time having to take the measure of to what degree Aristotelian philosophy was an appropriate intellectual tool for expressing Christian doctrine. (Ultimately, certain teachings of Aristotle were formally condemned by Church authorities, but Aquinas did not live to see it.) Aquinas was "all in" and, taking his cue from his mentor Albertus Magnus, fully integrated Aristotelian philosophy with Christian doctrine.

The switch from Platonic to Aristotelian categories was significant in its own right, as indicated by the difference in their respective theory of material reality. For Plato, individual material things are explained as copies of immaterial essences that take their reality by "participation"; by contrast, Aristotle developed a theory termed "hylo-morphism," in which the reality of individual material things is constituted by two elements: matter, which in its essence is chaotic and undefined; and immaterial form, which gives material things their physical shape and qualities and also provides a means by which their reality penetrates the immaterial mind of human beings. Aristotle's approach to the nature of physical reality gave it, as it were, more respect to physical reality than Plato's did; rather than being poor copies of ideas, material entities have an existence in themselves, which, by their form, could be understood and penetrated by the intellect.

Aristotle's more subtle and realistic metaphysics was reflected in the facticity of Aquinas's approach, which was not consonant with the more spiritual approach to Christian metaphysics and Christian worship that had hitherto influenced Christian thought. One no longer had to avoid looking at the material world as if it were simply a source of temptation to the believer, but rather as the field in which human beings live—and through and by means of which they must find their Christian salvation. Aquinas thus achieved penetration into such practical areas as philosophy of law, and within the realm of theology in his analysis of the body and blood of Christ during the mass when, according to Catholic doctrine, the material elements become fully penetrated by the divinity of Christ. The fine-tuned manner in which Aquinas could approach the difference and interplay between the material and spiritual realms of existence was derived in large part from his understanding of Aristotle's philosophy, on which he wrote several extended commentaries. As indicated in the previous chapter, the modern scientific approach would disjoin the Aristotelian intimacy between form and matter—between ideal existence and material existence—in its search for absolute control of physical existence by means of the application of totally abstract scientific laws.

Theologians have not lost the habit of utilizing contemporary philosophy to understand and express doctrine, as is exemplified in several twentieth-century examples. For even after the Middle Ages passed away and the modern world emerged, Christian and Jewish theologians continued to think and write in contemporary philosophical terms. Twentieth-century developments in European philosophy were dominated by analytic philosophy in England and phenomenology and existentialism in continental Europe, and each school has been utilized by theologians. Analytic philosophy, largely influenced by Ludwig Wittgenstein, has provided the intellectual context by which Anthony Kenney and others have reworked the thought of Thomas Aquinas, producing a school of "Analytic Thomists" who currently write philosophy and theology in a Christian key. The theological offspring of existentialism, largely influenced by Martin Heidegger, have been more numerous, including the Catholic thinker Gabriel Marcel, the prominent Protestant theologian Paul Tillich, and the Jewish sage, Martin Buber. The phenomenological school, founded by Edmund Husserl, may have been the most important provider of philosophic thought as a basis for religious thinking, for its offspring includes the ethicist Max Scheler (on whom Karyl Wojtyla, later Pope John Paul II, wrote a habilitation thesis); the Canadian Jesuit Bernard Lonergan; and Edith Stein, killed in Auschwitz, who was Husserl's assistant for a time.

Such a survey, however brief, of theological utilization of secular philosophy raises a question: since theologians and other religious philosophers have been so assiduous in attempting to explain Christian and Jewish doctrines in terms understandable by their contemporaries, why have they not attempted to utilize scientific thought in the same manner? Modern science may not necessarily be materialistic, atheistic, and reductive, but nonetheless it is a historical fact that with the rise of modern science, what had previously been a controversy about secular and revealed knowledge among theological academics has become a "steel-cage death match." After modern science enters the picture, it appears as if Athens, now represented by science, and Jerusalem, have nothing to say to each other.

While it is useful to understand the present debate surrounding evolution as the latest reflection of an age-old contest between secular and revealed knowledge, there does seem to be something different, more oppositive, in this latest incarnation. Perhaps the reason we sense this is that we are going through it, but our sense of an ultimate conflict between secular and revealed knowledge in the contemporary case is not only the result of loss of historical perspective; and the reason is that while philosophy often acknowledges the reality of spirit and the existence of God, science tends to be materialistic and atheistic. Thus, the twentieth-century schools of philosophy have not been able to provide a bridge, or a median concept, by which to connect science and religion. In effect, St. Augustine and St. Thomas had an easier time of it because both neo-Platonism and Aristotelianism are philosophies that acknowledge or attempt to prove the existence of one immaterial God, the reality of mind, and the immortality of the human soul—while modern science emphatically does not. Further, scientific explanation has traditionally tended to deny the legitimacy of the perception of meaning in the universe or of purpose-driven activity in its separate operations, pursuing a reductive agenda that has de-legitimized revealed knowledge. What, then, is the current state of relations between religion and science?

The "Two Track" Theory: Stephen Jay Gould's *NOMA*

One of Stephen Jay Gould's last books was *Rocks of Ages* (1999), whose tone is, if not elegiac, somewhat tired, for in this brief book, Gould gives us the benefit of a professional lifetime's effort of a Darwinian publicist and scientist contending with the religious enemies of Darwinian evolution.[4] It may seem odd, therefore, that in this short book, Gould praises no fewer than three popes, including Pius the Ninth, Pius the Twelfth, and John Paul the Second. There is a reason for this, however, for the latter two popes published documents permitting Catholics to research and accept some (but not all) of the main tenets of Darwinian evolution (i.e., Catholic doctrine excludes the Darwinian materialist thesis that all life is solely a mechanical process and asserts that all human beings have been provided by the Creator with an immortal soul). It appears that Gould set up these popes as a firewall against the objections of religious critics, including neo-Creationists and fundamentalists who oppose Darwinism as atheistic and unscriptural.

American public opinion has never accepted Darwinism wholeheartedly, and serious thinkers continue to make effective responses to it.[5] And while neo-Creationists may have lost the recent court cases, their argument that alternatives to Darwinism ought to be presented in high schools is an agreeable one to Americans. Gould probably feared that at some point, one of these cases is going to incorporate Justice Scalia's contention that American citizens ought to have a say in what their children are taught in public schools, thus sticking it to the Darwinians; and it is in this light that his book must be seen. Gould's proposed resolution of the religion-evolution controversy he calls "NOMA," an acronym for "non-overlapping magisteria." The basic idea is easily expressed: science and religion each occupy its own "frame" (a term borrowed from G. K. Chesterton), and each field should not exceed the proper limit of its frame. Gould willingly concedes that evolutionists have often overreached with declarations about religious matters, but his main concern is with religious believers who use revelation and the Bible to confute evolution.[6] The frames are important for Gould, since he perceives that if science and religion each stays within its own frame, there will be no further conflicts and neither side will be able to suppress the other. It sounds plausible, but religious believers who agreed to accept Gould's NOMA would be taking a sucker's deal, since he clearly expects that science will continue, as it has in the past, to confine religion into an ever narrower and more constricted frame while science expands its own frame into areas formerly occupied by religion.

Gould's NOMA theory is worth pursuing, because it is an explicit expression of the very popular but profoundly incorrect two-track theory about the proper relation between religion and science. Currently, influential scholars, including Michael Ruse and Francisco

Ayala, who deal with religion and science from an evolutionary point of view have apparently accepted it, and practically speaking, the two-track theory is the version of religion-science relations accepted by the prestigious American Academy for the Advancement of Science.[7] The two-track theory, it is to be assumed, provides a comfort zone—a "demilitarized zone"—between religion and science, such that each side can proceed in its own way without reference to what may be going on in the other. However, like real-life demilitarized zones that separate antagonistic borders, missiles and sabotage cross over them while each side claims the other is at fault. Legal cases brought against the teaching of evolution in public high schools, and other, more intellectual attacks on Darwinian evolution, have convinced the scientific community that they are under assault from elements that are descendants from Dark Ages. Meanwhile, religious believers remain wary of evolution and science in general; it is no use for avatars of religion-science amity (like Ayala, Ruse, Polkinghorne, and Gould) to assert that there is no inherent conflict between religion and science when proponents of Darwinian evolution (including, lately, Dennett and Dawkins; but previously, John Maynard-Smith, Herbert Spenser and T. H. Huxley) have claimed not only that there is a conflict, but that evolution is winning.

The surprising thing about the two-track theory is that it is not new, for, in the thirteenth century at the University of Paris, such a proposal was the thesis of a group of philosophers, including Siger deBrabant, who was accused of proposing the theory of the "double truth." In those days, the issue was not the Bible versus evolution—or science in general—but the tradition of revelation versus the newly discovered philosophy of Aristotle. A group of radical theologians, including Albert the Great and Thomas Aquinas, were attempting to integrate this newly discovered secular knowledge with Christian revelation and were facing heavy opposition from reactionary theologians. Siger's double-truth theory, like the two-track theory, was intended to alleviate the conflict that had become fierce and would eventually bring an ecclesiastical condemnation of Aristotelianism.[8] Thomas Aquinas, however, vehemently opposed the theory of the double truth with his famous dictum that all truth, secular and revealed, is from the Holy Spirit.[9]

Science versus Religion—a Draw

The present state of affairs in Western culture is that religion is in retreat as part of civil discourse—even in debates where religious input would be most helpful, such as human cloning or fetal research—while science and utilitarian ethics has seemingly captured the field. It might even seem that the tension between "secular and revealed knowledge" in its present form of "science versus religion" has been resolved; science has won, and religion is discounted as irrelevant, as a mere survivor from a less progressive time such as the Dark Ages or the 1950s. It must be admitted that there are good intellectual reasons, translatable into formal arguments, for opposition between modern science and revealed religion (just as there are good reasons to observe their deep commonalities). Science has its own implied metaphysics of Galilean atomism, which reduces physical reality to abstract mass points while discounting colors, motion, and other evidence of our senses as merely "secondary" qualities. Science has its uncompromising theory of causality, which combines materialism with mathematics so that the actions of bodies can be understood dualistically as contact and movement of basic particles on one hand and as the result of invisible forces described by calculus or probability on the other. Science also benefits in the latest version of the conflict from its own proclamations of impartiality, and from the putative superiority of its method, which supposedly produces at the end of its process a sure result, undeniable and irrefutable, so unlike theology and metaphysics. The triumph of the secular in our culture is, it is easy to assume, the direct result of the triumph of empirical science, and considering the formidable arsenal of scientistic arguments, it seems as if scientific secularism might have finally carried the day among Western intellectuals.

The triumph of science over religion, however, comes at a peculiar time, namely when science itself faces challenges to its cultural hegemony as never before since the Enlightenment, challenged not by a Romantic rejection of science's inability to comprehend humane values and religious context but by the denial of its very basis that science is the method of discovering ultimate truth. This challenge comes from college and university faculties, where postmodernist academics have developed entire schools of the denial of meaning in language, which promote the cultural relativity of truth, and who, after attacking the humanities and social sciences, are now aiming at the hard sciences. In this context, science is fighting for its academic life—for the oxygen of intellectual probity and the continued acknowledgement of its epistemological superiority that have provided scientists with approbation, authority, and funding. This postmodernist movement among the philosophers, *litterateurs*, historicists, sociologists, feminists, and multiculturalists is antiprogressive of course, more reactionary in its way than the theologians and Aristotelian philosophers who fought against Galileo, who at least believed that the universe could be understood by the human intellect.[10] Nonetheless, the left-wing attack constitutes an intellectual challenge that has not yet been successfully met by defenders of scientific objectivity.

The triumph of science has also been obstructed by developments from within science itself, since some of its basic theories, especially in physics, have developed beyond the simple-minded materialism characteristic of nineteenth-century thinking. Relativity theory and quantum physics propel us into levels of physical and methodological speculation so abstract that, according to philosopher of science Stephen Toulmin, philosophers and theologians have now reentered controversies about cosmology.[11] Physics, the base science, it now seems, can no longer give us visually precise pictures of either the atom with its myriad attendant particles and intermingling forces or of outer space, now filled with waves of gravity, black holes, and dark matter. A further effect is that pop culture now freely uses the terms of contemporary physics, such as "quantum jump," "expanding universe," "uncertainty principle," "anthropic principle" (in a novel by Updike), "event horizon" (the title of a movie) and "big bang" (also the name of a television series and, at one point, the name of a chicken sandwich offered by KFC). From within his own field of evolutionary biology, Gould was involved in sharp controversies surrounding determinism and chaos and was accused by other evolutionists of giving support to neo-Creationist deniers of Darwinism. The upshot of all of these developments is that, as John Polkinghorne has stated, in the arena of religion-science conflicts, "the days of the knock down argument are over"; which is to say that there are no longer the triumphant put-downs available that can be alleged as the proof that scientific reason must prevail over religious revelation, such as enabled Laplace to assure Emperor Napoleon that God was an unneeded hypothesis.[12]

Yet scientific secularism still prevails even as we are beyond the deployment of formal arguments in civilized contexts, as when Bertrand Russell debated Father Copleston on BBC about the existence of God. Today, instead, the formal "science vs. religion" debate has become so trivialized that the form was satirized by Monty Python and has degenerated to the point where a revival (underwritten by the Templeton Foundation) that took place in 1999 between two particle physicists was notable not for the deployment of further-refined arguments but for John Polkinghorne's sanguine assertion that religion and science are no longer in opposition and Stephen Weinberg's assertion that religion is "an insult to human dignity."[13] Not only is the day of the knockdown argument over, but it seems as if the day of *any* argument, in the formal sense, is over, which means that the science vs. religion controversy can no longer be settled decisively in intellectual terms; which means, further, that in the purely intellectual sense, science and religion have gained some form of parity.

But, then, in what terms can the present state of the science vs. religion controversy be understood, if not in intellectual terms? Ultimately, it seems as if the issue is not one of intellectual debate, since the scientists must now explain themselves in terms that are as abstract and arcane as those used in theology; are string theory and multiple universes any easier to explain than the doctrines of justification or the Trinity, and are they not as frankly distant from direct observation or experiment?[14] What has transpired so as to leave science triumphant despite ferocious questioning of its methodological legitimacy from left-wing academics and despite its recent turn to high abstractions amenable to philosophical and theological treatment? To answer this set of questions, we must distinguish scientific theory from it applications (i.e., science as explanation and science as technology), which is attempted in Chapter 14, on "techno-secularism."

Scientific Skepticism, Miracles, and Hume

One of the most peculiar and contradictory aspects of the development of modern science has been its connection to skepticism. Science, it is said, is skeptical, as if the whole purpose of scientific investigation is to throw doubt and to regard with suspicion every truth claim ever made by any human being. But the plain purpose of scientific research is to discover the truth about the nature of the constituents and causes that explain the physical universe. Skepticism especially attacks religious beliefs—and particularly, claims of miracles taking place and revelations to the faithful of certain ultimate truths that are beyond human understanding. Yet a consistent skepticism should attack not only religious truth claims but scientific claims as well. What is self-evident or accessible to our five senses of most of the grand theories that are put forward as representative triumphs of the scientific enterprise? The germ theory of disease is hardly self-evident, as proof of it requires the patient use of petri dishes to develop bacterial strains, the classification and identification of certain one-celled organisms under the microscope, clinical experimentation to connect certain symptoms to possession of identifiable one-celled organisms within the patients' organs or blood streams, and this process repeated many times so as to establish a general pattern of cause and effect. Yet the germ theory is at least amenable to sense experience under close and systematic observation; what to make, then, of the atomic theory of matter or Newton's law of gravity or super-abstract theories such as evolution and quantum mechanics? Skepticism, it would seem, should apply to such grand abstractions as much as the healing miracles of Christ, the revelations of the Law given to Moses, or the angel speaking to Joseph Smith.

Two consistent skeptics have appeared at either end of the scientific movement, Hume in the eighteenth century and Popper in the twentieth (although Popper, properly speaking, pursued a philosophy of "critical rationalism" rather than skepticism). Hume is the more interesting case, for Hume is often thought of as a scientific philosopher who applied the scientific point of view to religion. Yet this is not true, for what Hume brought to both religion and science was a general attitude of skepticism. What is true is that Hume, in an inconsistent manner, implied *sotto voce*—with a wink and a nudge—that what he was doing was demonstrating that miracles and revelation were impossible given the order of nature as determined by Newtonian-style scientific laws. Hume's skeptical arguments are famous among philosophers, so much so that what is overlooked is the inconsistency of his skepticism, for Hume's is a skepticism so all consuming that no intellectual vessel can hold it; it eats through everything and evaporates the intellect into an acidic mist.

The inherent contradiction of Hume's influential skepticism as it regards religion and science can be seen in his famous criticism of miracles, for here the basic themes of his philosophy coalesce into a system in which its power and its contradictions can clearly be seen. Hume states: "A miracle is a violation of the laws of nature, and as a firm and unalterable experience has established these laws, the proof against a miracle, from the very nature of the fact, is as entire as any argument from experience can possibly be imagined."[15] Stylistically, it is noticeable that Hume, while starting with an admirably clear definition of miracles, then adds a sequence of clauses, which tend to lead to confusion. He refers to "firm and unalterable experience" without indicating why any set of experiences could not be repeated in a fashion that in some manner contradicted the prior set of experiences. After all, Hume is famous for his destructive analysis of causality, in which he asserts that the detection of cause-and-effect relationships is in no way dependent upon reason but only on prior experience—and is, in effect, a mere habit of mind. "Were any object presented to us, and were we required to pronounce concerning the effect which will result from it without consulting past observations, after what manner, I beseech you, must the mind proceed in this operation? It must invent or imagine some event which it ascribed to the object as its effect; and it is plain that this invention must be entirely arbitrary."[16]

Now, when he refers to "past observations," Hume must be thinking, in effect, that the human mind is a *tabula rasa* as it faces each new kind of cause-and-effect phenomena, and thus any expectation of a causal relationship that is based on *a priori* reasoning is a matter of fantasy or invention. (The radical dichotomy between empirically validated statements and *a priori* statements is no longer taken for granted in empirical philosophy—see Quine's essay "Two Dogmas of Empiricism."[17]) To return to the original sentence in which Hume is treating miracles, the contradiction between the direct implication that laws of nature are immutable (as when he states "that a firm and unalterable experience has established these laws") and that the perception of causality can never be assured is so manifest that Hume apparently feels he must reinforce his point by means of the last clause, "the proof against a miracle, from the very nature of the fact, is as entire as any argument from experience can possibly be imagined." But is this so? If, as Hume famously asserted, it is possible to imagine a billiard ball evaporating into dust or flying up into the air, why can I not also imagine Joseph of Cupertino flying about his monastery or Muhammad flying on his horse to Jerusalem? The answer for Hume is based on direct experience, but why can there not be *new* experiences? Why must the order of nature be so determined that no expectation of miraculous events is possible? It is not likely that Hume, or any skeptical argument based on experience alone, can answer this question satisfactorily; but more to the point, what Hume is doing here is relying on an implicit determinism resting on Newton's mechanical view of the world. Hume's skeptical argument against miracles, as presented here, fails, since it is based entirely on experience—for why can there not be a new experience, for example, of a miracle? Only a doctrine of explicit materialism and determinism as provided by modern science can deny the possibility of a miracle *a priori*, which is, in effect, what Hume is attempting to do.

The detection of a basic contradiction in Hume's treatment of miracles is not new; the Cambridge philosopher C. D. Broad made the point in the last century.[18] What *is* important in this context is to realize that pervasive skepticism is not a necessary component of scientific method, and it is not implied as part of doing scientific research. That is, science does not proceed by generally denying the possibility that the human mind can know certain truths, as Hume does. Rather, the aim of science is to discover truths about the way the physical universe works and to explain types of physical phenomena in increasingly abstract and inclusive laws and theories. The triumphs of modern science have been its grandest theories—like general relativity, quantum mechanics, and evolution—but such grand theories are, in fact, mental constructs that transcend our experience of nature and are manifestly not provable by direct observation. Yet science must have them in order to make sense of a myriad of experimental phenomena and to provide comprehensive explanations; too rigid of a skepticism will kill off the flower of the scientific enterprise.

Yet scientists *are* skeptical, for in their search for causes, they eliminate, beforehand, accounts based on occult causes, such as voodoo; spiritual causes, such as prayer; or philosophical causes, such as design. In fact, only mechanical causes are considered as possible causes of any phenomena modern science studies, and the philosophy that derives from that is not skepticism but what is termed "methodological naturalism." But however successful the scientific method has been, as interested observers, we have the right to ask whether scientific method is the best or the only way to determine the truth about how things operate in the universe—*especially* those things that especially concern human existence. As we shall see, that depends on what kinds of matters science attempts to explain. We will examine that issue in some detail in Part III.

Notes

1. Frederick Copleston, S.J., *Medieval Philosophy*; New York, Harper Torchbooks, 1961;10.
2. Hart, Jeffrey, "Literature and the Foundations of the West;" *Modern Age*. Winter, 2000;63–71.
3. Copleston, 13, 14.

4. Stephen Jay Gould, *Rocks of Ages: Science and Religion in the Fullness of Life*. New York, Ballantine, 1999.
5. Michael Behe, *Darwin's Black Box*; New York, Free Press, 1997.
6. Gould, *Rocks*; 125–150.
7. The AAAS devotes significant resources to neutral discussions of the issues surrounding religion and science including publications and programs under the rubric, "AAAS Dialog on Science, Ethics and Religion" (DoSER), headed by its own director.
8. Copleston, 104, 5.
9. St. Thomas Aquinas, *On the Truth of the Catholic Faith: Summa Contra Gentiles* (trans. A. Pegis), Garden City, NJ, Doubleday Image Book, 1955; Book One, Chapter 7, "That the truth of reason is not opposed to the truth of the Christian Faith": 74–5.
10. See Gross, Paul and Levitt, Norman. 1998. *Higher Superstition: The Academic Left and Its Quarrels with Science*. Baltimore, Johns Hopkins U. Press.
11. Stephen Toulmin, *The Return to Cosmology: Postmodern Science and the Theology of Nature*. Berkeley, U. of California Press, 1982.
12. Polkinghorne comment . . .
13. *New York Times*. April 20, 1999: D-5. *The Chronicle of Higher Education* 1999; April 30, 1999; 17–18.
14. Wolfhardt Pannenberg, *An Introduction to Systematic Theology*. Grand Rapids, MI, Eerdmans, 1991; 37–52.
15. David Hume, *An Inquiry Concerning Human Understanding*; New York, Library of Liberal Ats, 1965; "Of Miracles;" 124.
16. Ibid., "Doubts Concerning the Understanding;" 43.
17. W. V. Quine, *From A Logical Point of View*; Cambridge, Harvard U. Press, 1961; "Two Dogmas of Empiricism;" 20–46.
18. C. D. Broad, "Hume's Theory of the Credibility of Miracles;" http://www.jstor.org/stable/4543915.

Chapter 7

The Counterrevolution in the Philosophy of Science

The purpose of this chapter is to describe a salient point in the evolution of the Western attitude toward modern science, as it has recently came about. The counterrevolution alluded to is a sea change, for contemporary thought has developed a well-wrought suspicion of the prior claim that modern science is a sure and certain source of human knowledge. But the rise of modern science provided, in effect, the intellectual justification of the Enlightenment, since it provided the sovereign proof of the ability of the human intellect to understand and control not only humanity's physical environment but also its social and mental environments. And with the sea change in our attitude toward the epistemological guarantees of science comes a suspicion that our confidence in humanity's powers to control its destiny, as well as its environment, is no longer justified; the historical record of the twentieth century seems to provide enough evidence to show that this general decline of confidence is justified. But prior to describing this cultural sea change and how it came about, some background about an early-twentieth-century intellectual movement is required.

The Unity of Science Movement

The Unity of Science movement developed in the earlier part of the twentieth century in Central Europe but eventually, under the pressures of war and dislocation, spread to England and America. The movement itself was the offshoot of the famous Vienna Circle, a collection of intellectuals and thinkers the central concerns of which were twofold: first, to locate the precise boundaries between real knowledge (which was assumed to be available only from empirical science) and false claims to knowledge (i.e., "metaphysics") and second, to establish a social environment that reflected the movement's belief set (in effect, a quasi-socialist integration of education, economics, and social concern conformable with the standards of scientific knowledge). At that time in Central Europe, there were many powerful intellectual movements developing, associated with Einstein's relativity theories, new discoveries in mathematics and logic, and developments in pedagogy and empirical psychology. The Vienna Circle was formed and led by Moritz Schlick, a professor of philosophy, and was made up of philosophers, logicians, mathematicians, and economists. The basic, commonly held philosophy was positivistic, based on the premise that empirical science alone was the true source of knowledge, which led them, in figurative terms, to follow Hume's advice, throwing any book or doctrine that could not be scientifically verified "into the flames." This meant sharply distinguishing science from metaphysics, including speculative philosophy and religion, and while the impulses that directed the activities of the Vienna Circle arose mainly from new and exciting developments in all the fields of science, the Vienna Circle program reflected earlier forms of positivism and was, in actuality, a full-blown example of "scientism."

The Unity of Science movement was a result of this and was the brainchild of the Circle's most energetic personality, Otto Neurath. The Circle, however disintegrated under the pressures of Nazi rule, and many of its members and associates fled to England and America; the émigrés included Popper, Carnap, and also Neurath. Resembling what happened to Judaism after the destruction of Jerusalem, the philosophical project of the Vienna Circle, having lost its geographical base, became a matter of interpretation of texts. In the case of the Circle, the texts were not the books of the Hebrew Bible, but rather, a project—the Unity of Science movement, which aimed at composing an international "Encyclopedia of Unified Science." Like the French encyclopedia of the eighteenth century, this encyclopedia would be written by many scholars and not be the work of one man. Volumes 1 and 2 of *The Encyclopedia of Unified Science* were eventually published—not in Vienna, however, but in Chicago by the University of Chicago Press—and included book-length works by Rudolph Carnap, the logician, and Charles Morris, the linguistic theorist.[1]

Carnap perhaps is the most representative of the authors who were members of the Vienna Circle who contributed to the movement; his best-known work, *Die Logische Aufbau die Welt,* laid out the process by which scientific knowledge arose from a basic level of empirical description enunciated in "protocol statements," which then proceeded to theoretical or general statements.[2] In this way, Carnap was expressing in the most formalistic way possible (i.e., by means of logical expressions) the basic tenet of positivism, for the underlying premise is that only empirical statements and scientific generalizations provide knowledge. But in explaining what Carnap called the logical "construction" of the world, he was also showing a subtle development in positivistic thinking, one that shows a slightly postmodern tendency: namely, a reflection upon the fact that human beings make a logical construction out of their own experience, so that, epistemologically, what is taking place is not an immediate knowledge of the external world but, in some sense, an effort by human minds to know the environment in which they live, move, and have their being, a tendency that reflected the influence of Mach and Kant.

Another concern of the Vienna Circle thinkers was the social aspect of the process of scientific knowledge, which was understood to involve the whole community of man, unlike the Cartesian process, in which certainty is reached by one individual searching the inner precincts of his own mind. Despite these epistemological subtleties, however, the point of the Circle and the Unity of Science movement remained that scientific knowledge was the definition of true knowledge, and so the forbidding sentences of Carnap's logical expressions implied that empirical science was the key to knowledge that provided the rational basis of social policy. Tradition, speculative philosophy, and, certainly, religious belief were not acknowledged as anything but "metaphysics"—used here as a pejorative term that implied that any rational and up-to-date person would reject such unworthy opinions out of hand. The thought of the Vienna Circle was brought to England by A. J. Ayer, who studied in Europe as a young man and who, once he returned to England, wrote *Language, Truth and Logic*, a vivid exposition of the positivistic philosophy of the Circle.[3]

But having acknowledged the subtlety of the process of scientific knowledge, and having been influenced by the current examples of revolutionary developments in science—including relativity and quantum physics—how sure and certain could scientific knowledge actually be? After all, the revolutions in physics going on at the same time as the Vienna Circle thinkers were working indicated that a long-established theory, Newtonian mechanics—the most influential and longest-standing theory in the history of science—was in fact false, or at least incomplete. It was the breakup of the grand synthesis of Newtonian mechanics that had dominated scientific thinking for two centuries, guiding it to unparalleled and undreamed-of success on both the level of theory and practical application, that was the main motive for rethinking the basis of scientific knowledge. Einstein particularly was an exemplar (not to say *hero*) of the Vienna Circle thinkers; Karl Popper made it a point to correspond with Einstein, while Einstein himself was celebrated as much as a philosopher as a scientist.[4]

From the beginning of the movement, a distinct problem had arisen: namely, that given the ever-present example of the breakup of the Newtonian system, scientific knowledge could never be understood to be certain; but on the other hand, scientific knowledge was held to be the only source of true knowledge. But then, if science was to be distinguished from metaphysics, what was the distinct character of scientific knowledge that distinguished science from metaphysics? Thus arose the so-called problem of *demarcation*. One possible solution was to say that *all* knowledge was uncertain, and in fact, the logic of the demarcation problem seemed to push the Vienna Circle thinkers into a position of epistemological relativism. However, one thinker who was not a recognized member of the Circle but nonetheless closely associated with it was Karl Popper; Popper refused to follow the logic that led to relativism and devised a criterion that served to distinguish scientific theories from nonscientific statements. Popper's solution was *falsification*, which meant that truly scientific statements were expressed in such a way that they could, upon empirical evidence, be shown to be false; however, given the intrinsic uncertainty of scientific knowledge from a historical point of view, no statement or assertion could ever be shown to be true in all times and in all places. Statements or theories that could not be falsified, Popper said, were, in the term of dismissal used by Vienna Circle theorists, mere "metaphysics." So by means of Popper's falsifiability criterion, the distinction between science and metaphysics, including political theory, philosophy, and religion, still held—but for how long?[5]

Popper's rational, critical approach was not merely a philosophical position but was strongly reflective of his personal attitude as well, for this constantly critical approach was not merely a personal crotchet but had a philosophical resonance. Popper seemed to have a spiteful, dog-in-the-manger attitude about scientific knowledge, as if to say "If, as the history of science seems to show, it is not as Bacon and the nineteenth-century positivists thought, that modern science is a sure and certain, and the only secure, source of knowledge, then no other human activity can be a source of knowledge, either." In this, Popper's philosophy resembles that of the great skeptic Hume, except that unlike Hume, Popper retained the ideal of science as true knowledge, an ideal he could never give up—which seemed to make him perpetually frustrated and angry.

Like Hume as well, this skeptical attitude had consequences for Popper's political philosophy, expressed in his famous book *The Open Society and Its Enemies*, for he was skeptical of the claims of revolutionaries and ideologues who wanted to reform society according to a rational plan or to transform human history by means of a cadre of disciplined believers. He aimed his sharply critical approach not only at Plato and Hegel but also at Marx, to whom he devoted most of the second of two volumes of *The Open Society*, which made him a premature anti-Communist and which served to alienate him from the dominant left-wing intellectual culture, casting him into the circle of the excluded along with Koestler, Malraux, and Chambers.[6] Popper's politics was not conservative but, properly speaking, a libertarian philosophy, one that would later accord well with the major themes of the Thatcher government (Popper was given a knighthood for his intellectual support).

Reading Philosophy of Science Prior to the Counterrevolution

It is an unnoticed irony in the history of the Vienna Circle and the Unity of Science movement that the final contribution they brought forth, the last and culminating book of the *Encyclopedia of Unified Science*, was Thomas Kuhn's *The Structure of Scientific Revolutions*. To show why that was, it is useful to point out how Kuhn's account of science, developed from studying its history rather than its logic, contradicted the thought of the Vienna Circle, which had previously dominated the field of the philosophy of science.

The basis of most of the philosophy of science prior to the counterrevolution was empiricist and effectively positivistic, for the major influences such as Carnap and Phillip Frank had come across from Europe and established themselves in the United States and England. Nonetheless, the basic concern of some students of the philosophy of science (such as the present author) was to try to understand modern empirical science in a way that was agreeable with humanistic understanding; in other words, to overcome the positivistic doctrine that science represented some sort of true knowledge while what was relegated to metaphysics was mere myth, at best a "propaedeutic" (in Marx Wartofsky's phrase), which could serve as a preliminary attempt at a valid scientific theory. A representative example of the state of the field prior to Kuhn's historical philosophizing, and also that of Hanson, Lakatos, Feyerabend, and Polanyi, was the volume entitled *Readings in the Philosophy of Science.*

Readings is a formidable volume that runs close to eight hundred pages, filled with articles from the foremost scientific philosophers of the day, including Carnap, Russell, and Einstein. The readings are empiricist and positivistic in a manner that severely restricts the least attempt to define something outside of science as either important or interesting. Its point of view is well summarized in an introductory chapter by May Brodbeck, one of the editors, who wrote: "Modern philosophy of science . . . is a specialized part of analytical *philosophy.* Modern analytical philosophy is unlike traditional metaphysics in its rejection of the latter's implicit assumption that it is the philosopher's task to give supra-scientific descriptions and explanations of the nature of reality." To drive home her point and to make her outright hostility apparent, Brodbeck further wrote, "Cosmological speculations about the origin, nature, and purpose of the universe may be harmless, imaginative extrapolations from science itself. But the philosopher of nature's vaunted metaphysical explanations of the universe, allegedly cutting deeper than mere empirical scientific explanation, were so patently anthropomorphic and metaphorical as to beg all the issues that philosophers must analyze."[7] All in all, *Readings* did not make doing philosophy of science look like much fun.

Another prominent textbook of the time, which extended the positivistic understanding of science to the social sciences, was *The Conduct of Inquiry* by Abraham Kaplan, which neatly divided up the topic into several general categories and subcategories—for example, "methodology," "models," "theories," and "values"—and treated the philosophy of social science with seeming comprehension.[8] Yet there were no sections or subsections on the separate social sciences such as psychology, sociology, anthropology, economics, and so on; thus the book made no effective distinction between the various fields, as if to imply that the methods of, for example, psychology are the same as those of anthropology or economics. Also, the textbook never considered basic questions such as "given the success of modern science in prediction and control of phenomena, how can this be explained?"

That is, what is the "fit" between the human intellect and the physical universe that science investigates? What is the significance of the fact that experiment and observation are systematically utilized by modern science to produce comprehensive theories of a range and application that would have astonished Plato? More pertinently for the philosophy of social science, what are the implications or consequences of applying scientific method to the study of human nature, and what, if anything, can social science inform us of in general terms about human reality that is not known from religion, art, literature, politics, history, and philosophy? And what about "values" and religious beliefs? How should social science evaluate these? Indeed, should science take them into consideration at all? Alas, these points, and such others, were not discussed in Kaplan's book.

Kaplan's general philosophical approach was "phenomenalistic," which needs to be distinguished from "phenomenological." The difference between the two words is significant, for it reflects two very different approaches to the philosophy of science (and philosophy in general) as it has developed in the twentieth and now the twenty-first centuries. The phenomenological movement is continental, stemming largely from the teaching of Edmund Husserl, which has extended its influence into ethics and social science; one of its offspring is the philosophical movement known as existentialism. The phenomenalism that characterizes Kaplan's approach depends on the English (and now American) tradition of empiricist philosophy, which looks to the immediate elements of sensory experience—that were termed "ideas" by Hume and "sense data" by Russell—as the basis of knowledge. Since it is possible to examine each immediate sensory experience as separate and isolated from any other sense data, the problem of knowledge then becomes one of assembling a bundle of separate sense data into a single substance, such as an apple. This is an impossible task—that is, of giving a clear description of how this process takes place and then, more importantly, of providing a justification for the validity of knowledge achieved in this manner. This kind of phenomenalism was the basis of the philosophy of science done at that time (G. E. M. Anscombe, the influential British philosopher, was devoted to Ludwig Wittgenstein, because his method had relieved her of her belief in phenomenalism.)

Kaplan's book, however, was typical of other major monographs in the philosophy of science in those days, of which *The Structure of Science* by Ernest Nagel was the chief representative. It, too, divided up the subject of philosophy of science into general categories and discussed each of them in a dry, empiricist manner that left out the large questions mentioned above, except to dismiss them as unwarranted objections—as in the case of whether there is a difference between studying social and physical phenomena (Nagel's answer was "no").[9] This book runs six hundred pages of text and remains an impressive monument to the philosophy of science done in the 1950s. It was comprehensive in that it dealt with all of the sciences—including physics, biology, and social science—but its basic tendency implied that physics was the base science, ontologically speaking, which provided the model of how science should be done. As a result, there was nothing distinctive about problems in biological or social science that, in any essential way, distinguished these scientific fields from physics; the chief means of making this point was Nagel's championing of a complete reductionism.[10] The general impression left by this comprehensive, but now out-of-date work was that there were no unresolved—or interesting—issues left in the philosophical understanding of modern science. One prominent philosopher of science said that Nagel's book resembled a steamroller leaving no sign of philosophical life behind it.

Kuhn's Famous Book: From Positivism to Historicism

Kuhn's book *The Structure of Scientific Revolutions* is well known for his description of the course of the history of science as taking place in a succession of "revolutions" that cause "paradigm shifts," of which the greatest example was the Copernican revolution (on which Kuhn had published a previous book), but also, by implication, the overturning of the Newtonian mechanical worldview by relativity and quantum theory. The paradigms themselves, established by the so-called revolutions, were not merely large theories but also had a social coherence that included a sense of membership among those scientists who accepted them and which also included the typical ways of doing experiments and providing a template of the range of acceptable explanations for the phenomena being studied in a particular field. This social turn reflected the kind of thinking that was already part of the Vienna Circle approach, that is, that science is a social enterprise; however, Kuhn expanded and intensified this social turn to account for the manner in which scientists accepted their paradigms, which he claimed depended not so much on empirical evidence and inductive logic but on the sense of communal identity and personal commitment.

In using the history of science rather than inductive logic as the basis for understanding modern science, Kuhn's theory came in direct conflict with the major thesis of Vienna Circle positivism, namely that there is a *demarcation* between science and metaphysics. In effect, Kuhn's thesis implied that there was no demarcation between science and other fields of human endeavor—including, specifically, politics, for he made direct comparisons between revolutions in science history and political revolutions.[11] Kuhn's thesis even implied that there are specific resemblances between science and religious belief in his descriptions of what it is like for scientists to accept a paradigm, for he asserted that scientists, in point of fact, did not give up the theories to which they were committed when contradictory evidence appeared; instead, they simply made up additional hypotheses to explain the contradiction or, in extreme instances, asserted that what appeared to be contrary evidence did not count as evidence. (The classic example is when medieval astronomers invented deferents and epicycles to account for the contradiction between Earth-centered astronomy and increasingly accurate astronomical observations.) It was when an overwhelming number of examples of contrary evidence appeared that the accumulated conflicts would force the creation of a new paradigm. But Kuhn's account of paradigm commitment among scientists eerily resembles religious commitment, as when believers continue to believe even in the face of contradictory evidence, such as death and disaster in the face of belief in an all-powerful and loving God.

Kuhn's account could effect a repair in the breach between science and the humanities, for his account of science made it understandable in historical terms. And in reducing the line of demarcation between science on one hand and philosophy, politics, and religion on the other, Kuhn made it possible for readers without a technical knowledge of physics, chemistry, or biology to approach modern science in the belief that they could understand its essence. Since Kuhn's time (and reflective, in part, of his influence), the academic study of science has changed focus from the philosophy to the history of science. Thus, two academic institutions established for the philosophy of science in the 1960s at Indiana University and Boston University have since changed their focus to the philosophy and history of science. Meanwhile, the field called *science studies* has arisen, which applies sociological and historical methods and explanations to account for specific episodes in scientific discovery. This is, in essence, what justifies calling post-Kuhnian study of science a sea change or the result of a counter-revolution, for Kuhn's theory of paradigms that elevated the philosophy of science beyond a study of its epistemology and method, which had positivistic and reductive consequences, and looked rather to the historical and social consequences of modern empirical science—modern science seen not as the inevitable consequence of the workings of human reason but as a cultural event. In the broadest terms, the difference between pre- and post-Kuhnian philosophy of science is the difference between positivism and historicism.

While Kuhn's book was met with appreciation from social scientists, including psychologists, sociologists, and historians, it met with severe criticism from Kuhn's colleagues in the philosophy of science. Two examples suffice: First, Kuhn's thesis had undergone a barrage of criticism from other philosophers, which had forced him, in a second edition, to enlarge his original text by adding a thirty-six-page postscript to extend his remarks. Second, a book, *Criticism and the Growth of Knowledge*, appeared, which was a report of a conference held in Bedford College, London, in 1965, which began with a longish paper by Kuhn that was criticized by a succession of eight critics who included the best-known and sharpest minds in the philosophy of science field, including Toulmin, Feyerabend, Lakatos, and, of course, Popper. At the end of the book, Kuhn added another long essay, "Reflections on My Critics."[12]

It may be said that Kuhn's book itself made a revolution in the philosophy of science, but such a view gives a false impression—as if the independent actions of Washington started the American Revolution as its sole cause, as if there were no Adams, Jefferson, Franklin, Hamilton, or Light Horse Harry Lee involved. There is always a social context to revolutions, and on reconsiderations, such sudden events, no matter how climactic they may appear to be, are less likely to be characterized as revolutions than as *evolutions*. It is very much the case with Kuhn, as he was representative of a whole movement by philosophers, historians, and scientists themselves—as they, as a group, considered the effects of the revolutions in science that had taken place in relativity, quantum physics, and mathematical logic. Einstein and Heisenberg, as well as other scientists, considered the research they had done and concluded that the old, Baconian account of how scientific knowledge is attained was faulty, since it was, broadly speaking, not the case that scientific laws and theories were generated as a direct result of studying and mathematically categorizing empirical data but that rather, that research began with a theory or an idea of how phenomena could be explained as a guide to what empirical questions would be asked; in short, theory was prior to the collection of evidence in science.

Among the philosophers who were taking a new view of science, several deserve mention. Norwood Russell Hanson was an American studying at Cambridge University after World War II, who wrote *Patterns of Discovery*, which introduced the expression "theory-laden observation terms." Hanson returned to the United States, where he set up the Philosophy of Science program at Indiana University, which was one of the very first such programs in America. Hanson died when the small plane he was piloting crashed; he had been a pilot in the war. Steven Toulmin studied under Wittgenstein at Cambridge University and applied his mentor's antinomian philosophy to the study of science. Toulmin produced, among a number of other works, two devoted to the philosophy of science: a short textbook, *Philosophy of Science*, and later on, *Cosmopolis: The Hidden Agenda of Modernity*, a book that explained and furthered interest in the new development of cosmological thought in science. Polanyi, a Hungarian chemist, worked for many years developing an intricate argument and presentation about how scientists actually come to know what they know; *Personal Knowledge*'s emphasis was not on the attainment of abstract theory but on the more intimate details of how scientists, and human beings in general, came to acquire knowledge.

More closely related to Kuhn's ideas were those of Imre Lakatos, an eastern European who succeeded Popper in his chair at the London School of Economics. Lakatos revised Kuhn's idea of paradigms, calling them *research programmes* [sic], putting more emphasis on their explanatory aspect than Kuhn had done. Thus, Lakatos stated that there are "belts" of explanatory terms and theories that surround the basic theory of what Kuhn called "paradigms" that, as it were, protect it from contrary evidence and arguments. In this way, he explained how it was that basic theories could maintain interest among scientists and guide research even in the face of apparent evidentiary contradiction. Also connected to Kuhn, albeit in a somewhat more personal way, was Kuhn's colleague at the University of California, Paul Feyerabend. Feyerabend was a ferocious contrarian and self-appointed gadfly who espoused epistemological relativism in philosophy and anarchy in politics. His most well known work is *Against Method*, in which, among other things, he argued that the Catholic Church had acted properly in silencing Galileo. He argued persistently with Kuhn when they were colleagues; he claimed that Kuhn had once thrown an ashtray at him.

Two other well-known and influential philosophers whose thought should be understood in relation to Kuhn's—and to the counterrevolution in the philosophy of science—are Karl Popper and the ethicist Alasdair MacIntyre. Popper has already been mentioned, but the point here is that he is a transitional figure between the positivism of the Vienna Circle theorists and the historicist understanding of science that has largely succeeded it (one of his shorter works is *The Poverty of Historicism*). His attempt to formulate a philosophy that would retain the ideal of science as the true form of human knowledge, while at the same time making allowances for the fact of scientific error and disproof, manifested a brilliant mind attempting to do the intellectually impossible; for if science could not be distinguished as the one source of truth among all the other contenders—from literature and myth to politics and history—because of its inerrancy, then why should science retain its social authority? Popper's theory that the falsifiability criterion could make the distinction also implied that what scientists did was to try to invent non-falsifiable theories, rather than act from the positive intent to discover the truth about the physical universe.

Perhaps the most influential ethicist writing in the English language in the past twenty years is Alasdair MacIntyre, who is credited with having added a new focus to the theory of ethics beyond the usual categories of utilitarianism, intuitionism, duty-based ethics, and emotivism. In the 1970s, MacIntyre was hired by Boston University's controversial president John Silber to head up the Philosophy Department. The university at that time had a robust program in the philosophy of science, while MacIntyre had established a reputation as a historian of ethical theory and a commentator on contemporary culture (e.g., in *A Short History of Ethics*). While chairman, MacIntyre was introduced to the controversies then swirling about the philosophy of science—in particular, the ideas of Kuhn and Lakatos, which he credited in the preface to his most famous work, *After Virtue*. It seems as if MacIntyre was at a standstill in his philosophical understanding of the course of ethical ideas in Western culture, and the new historicist ideas that placed emphasis on the social and cultural aspects of scientific explanation affected his thought deeply, giving MacIntyre a new emphasis in the understanding of the evolution of ethical ideas in the West. After *After Virtue*, MacIntyre continued his thinking and writing along the same historicist lines in several consequent volumes, developing a new direction in ethical theory, *virtue ethics.*

The Politics of Paradigms

While some critics were concerned certainly with questions about the logic of scientific method expressed in Kuhn's theory, others strikingly were more concerned with what they perceived as Kuhn's reactionary political position. Politics of the common sort were not at issue, as between the Labor Party and the Tories in England, or Republicans and Democrats in the United States. Rather, as Feyerabend and Popper stated, it was that Kuhn's theory, in effect, denied the great founding myth of modern scientific philosophy, namely that the development of modern science was an essential and identifying element in the Enlightenment progress that brought about the modern age of free thought, technological development, and political liberty. Kuhn's theory of paradigms implied that scientific progress was not based on an accumulation of newly discovered scientific truths but was a process of scientists simply changing their minds, less because evidence impelled the change than because the general social context—"mood" might express it—was no longer the same. Kuhn, it is important to note, denied the charge of epistemological relativism, which was the main reason he was always being forced to reply to critics who kept making the charge.[13]

The connection and the political implications of Kuhn's theory, while political, were not partisan. The connections were political in the deepest sense, for what the controversy and the criticisms were about was twofold. First, it was a personal issue for some of the critics, for it now appeared as though the philosophy of science was hitherto to be dominated by a new approach based primarily on the history of science. In the past, concentration had been on the methodology of science, and historical events were brought up to illustrate the point being made by the philosopher. Kuhn, however, had, like some continental philosophers, based his philosophy on the discernible patterns of discovery and invention in the history of science. Kuhn, it seemed to me, had more in common with philosophers of history such as Vico and Hegel than he did with Bacon and Russell, a fact that made the controversies surrounding his account confusing both to his critics and possibly to Kuhn himself.

The major point at issue concerned the Enlightenment, for to have, as it were, historicized the understanding of science was also to have relativized science as a source of knowledge, which in turn cast doubt on the Enlightenment conceit that human rationality, once freed from medieval oppression, fostered a new age of constant progress. This new scientific age would be humanistic rather than religious at its basic understanding of human life and universal history—from St. Augustine to H. G. Wells, so to speak. Since its publication, Kuhn's thesis has had its effect, observable in the fact that philosophy of science is no longer a "hot" field and has, in effect, been replaced by the history of science as the principle means of understanding the nature of modern science and its role in the general culture of the West and the world. Books are now written that show science in its cultural emplacement so that it becomes possible, for example, to seriously discuss whether the Church was correct in its rejection of the Copernican thesis during the Galileo controversy or whether German scientists during World War II deliberately sabotaged Hitler's desire for the atomic bomb.[14]

Notes

1. The second edition of Thomas Kuhn's *The Structure of Scientific Revolutions* (Chicago, U. of Chicago Press, 1970) contains references to the Encyclopedia on the title page and on pp. ii, iii.
2. *Die Logische Aufbau Die Welt* (1928) in translation as *The Logical Structure of the World and Pseudoproblems in Philosophy* (trans. R. A. George);Open Court, 2003.
3. Frederick Ayer, *Language, Truth and Logic*; New York, Dover, no date.
4. See Paul A. Schillp, *Albert Einstein, Philosopher Scientist* (2 vols); LaSalle, Open Court, 1970.
5. Karl Popper, *The Logic of Scientific Discovery*; London, Routledge, 1992; 57–74.
6. Karl Popper, *The Open Society and Its Enemies*, Vol 2; Princeton, Princeton U. Press, 1966.
7. Herbert Feigl and May Brodbeck, editors, *Readings in the Philosophy of Science*; New York, Appleton-Century Crofts, 1953; 5.
8. Abraham Kaplan, *The Conduct of Inquiry*; SanFrancisco, Chandler, 1964.
9. Ernest Nagel, *The Structure of Science* ; New York, Harcourt, Brace and World, 1961.
10. Ibid., Chp. 11, "The Reduction of Theories;" 336–397.
11. Kuhn, *Revolutions*, on political paradigms; 94.
12. *Criticism and the Growth of Knowledge*, edited by Imre Lakatos and Alan Musgrave; Cambridge, Cambridge U. Press, 1970.
13. Kuhn, *Revolutions*; "Revolutions and Relativism"; 205–207.
14. On Galileo, see Paul Feyerabend, *Against Method*; New York, Verso, 1988; 129–138.On the German A-bomb, see a popular summary by Lee Kryzstek, "Did the Nazi's Build an Atomic Bomb?"; The Museum of UnNatural [*sic*] History; http://www.unmuseum.org/nbomb.htm (accessed 12-10-2012).

Part III

Science as Cultural Unifier: Three American Scientists

Chapter 8

Atoms in the Cultural Void: Steven Weinberg's Material Dreams

This part of the book consists of three chapters analyzing the effects, intended and unintended, of contemporary empirical science on American culture. Following the tripartite division of the sciences described in the fifth chapter, this part will analyze physical science, biology, and social science as culture formers by concentrating on three representative American figures, Steven Weinberg (atomism), Edward O. Wilson (social biology), and Stanley Milgram (scientific authority).

Physics and Politics

From a perhaps recherché perspective, the development of modern science resembles a revolutionary movement to capture the soul of the West. A power base is established first, which is the deployment of the physical sciences of astronomy, mechanics, and chemistry, displacing not only the old physics of Aristotle (his theory of motion and impetus) but, by implication, the old metaphysics as well. The perspective of modern science as a revolutionary movement may not be as recherché as it seems, however, for it is a historical fact that reduction, which is the leading edge of the scientific capture of a given area of human experience, has proceeded in three identifiable stages. Furthermore, this understanding of reduction as revolution provides not only a historical perspective on the development of modern science itself but also on the development of politics and modern culture at large.

The first of the three major fields of modern science to develop was physical science; and the social, cultural, philosophical, and even religious implications were immediately apparent. Indeed, the onrush of the new science, in all its aspects, proceeded along so many fronts that its ultimate significance goes well beyond the arrival of a new method for empirical research. It appeared to be a wholly new form of knowledge that gave to the human race control of its physical environment, and therefore the arrival of science constitutes a major cultural event. More precisely, science was an essential element in the Renaissance effort to put mankind in control of his own destiny, apart from the influence of religion, the state, and the oversight of God. Thus, while the new physics concerned mechanics and astronomy at first, the purpose of this research activity was not disinterested knowledge, even though the researchers themselves may have thought so; it ultimately aimed at rational control of all aspects of human experience and physical reality—in fact, of the whole universe, interiorly and exteriorly. Historian Peter Gay noted of the French *philosophes* who were the most consistent interpreters of the new physics in cultural terms, "They pushed its philosophical implications far beyond what the scientists themselves would have thought warranted. They tried to apply the scientific style of thinking to the regions of aesthetics, social and political theory. But they discovered that, having eliminated the problem of God, they had burdened themselves with new difficulties."[1]

The new theories in astronomy and motion had an impact in political theory, as seen in the development of the contract theory of government. The connection between the contract theory of government and the models of eighteenth-century physical science may not seem obvious, but it is there. The point at which to begin is to consider this question: how could the model of a contract be thought useful as the ultimate analysis of government and society—and of the relationship of an individual to his society? Consider the objection to using a contract for such a purpose: In a contract, representatives of two entities (persons or corporate bodies) agree to exchange service for a consideration (e.g., money). But the conception of a society made up of a number of such individual contracts (one for each citizen) is not a literal description of what actually takes place between a person and his society; nor, despite its influence and popularity, is the analogy a good. Consider the almost-obvious objections.

First, contracts are *agreed* upon, whereas one's loyalty to one's family, city, or nation is taken for granted, so much so that disloyalty is spoken of in the most pejorative terms as *treason*. (In the eighteenth century, "high treason" referred to the disloyalty of a person to his king, while "low treason" referred to the murder of a husband by his wife.)

In fact, no explicit agreement is expressed or implied. As someone's son or daughter, a person simply owes devotion; as a freeborn citizen of one's nation, one is simply expected to have an allegiance to it. Contracts imply freedom and consent, but one is not "free" to be born to one's parents, nor does one consent to be born a citizen of one's nation or state.

Another objection is that the contract theory implies a tension between the individual and one's social environment. Such tension is factually present in many cases, as expressed in the ancient Greek plays or during times of revolution. But even as tension between self and society is often present to some degree, the contract theory makes it constant and inevitable, so much so that a false dichotomy appears, and problems of the individual and the state's authority are unnecessarily intensified. The connection, therefore, between contracts and the place of the individual in society is not altogether an obvious or natural one, such that a philosopher can erect an entire theory of government upon it without extensive elaboration (most recently in Rawls's long and very-detailed volume *A Theory of Justice*.)

Then there is the nature of the individual him- or herself—although those pronouns are rather too specific; the individual *itself* is more appropriate, for in the contract view, the individual has more the qualities of an atom than a person. Fatherhood and femininity are not the qualities that Hobbes details in his description of the atomic individual within his mechanical view of human society; rather, dominance and fear are among the chief characteristics of human nature in Hobbes' view, and later theorists depending on the contract view of the atomic individual ascribe, as economic theorists tend to do, rational self-interest as their chief characteristic. In these examples, and more generally in attitudes dominant in society today, the individual is set far apart from its society such that the inherent need and desire of individual human beings for the consort of other people and a place in social setting is not considered as an inherent or essential fact. Thus, the ties that bind individuals to one another and to society at large become understood as rationalistic constructions or, as in the case of Hobbes and modern totalitarian practice, the imposition of the fear of death in order to provide for social cohesion.

From a methodological point of view, what has taken place closely resembles the scientific process of reduction. Just as it is typical of the method of mechanics to reduce complex motions to a combination of simpler ones (and Galileo's theories about motion influenced Hobbes in this regard) and to conceive of physical bodies as congeries of atoms, so also what Hobbes and Locke and other contract theorists are doing is reducing society to the actions of its atomic constituents (i.e., to the actions of individual citizens). From this point of view, the origins of the contract theory of government are as much to be seen in the newly developing physical science of the time as in the newly developing mercantile economy. Hobbes based his idea of the state on the contract theory not because he was committed to the idea of individualism, but because he was modeling his analysis on the reductive method of science. Later contract theorists may seem to have avoided this connection, but it is relevant that Locke, for example, was a good friend of Isaac Newton's, although here, the influence of science is not in the immediate application of reductive physics but in a more generalized sense of scientific rationality.[2]

Weinberg on Cosmic Origins, Science, and God

No contemporary physicist knows more about atomic structure or is more conversant with cultural implications than Steven Weinberg. He is a Nobel laureate in physics, the author of scientific papers of exceeding abstractness, and the writer of well-received expositions of quantum physics and its implications for a popular audience. Weinberg's best-known, semi-popular work is his book *The First Three Minutes*, which is a detailed description of the big bang and the immediate aftermath that described the mist of particles, quanta, high energy, and other forces that emerged in such a way at the moment of the big bang that distinctions between types of particles and types of energy are of no use.[3]

Weinberg also explains that in the view of contemporary physics, which treats the following as a settled doctrine, there are four basic forces that form the universe: the strong nuclear force, the weak nuclear force, electromagnetism, and gravity. As hinted at above, the strong nuclear force binds the internal parts of atoms together; it is the sundering of this very powerful force that explains the power of the atomic bomb, which, by splitting uranium isotopes, releases their interior energy. Then there is the weak nuclear force, which allows the atoms to bind with other atoms, in effect allowing electrons that, while they circle the nucleus, are detachable from it; the combining or sharing of electron shells accounts for the fact that atoms can "stick together" to form molecules and complex compounds. While the two nuclear forces are undetectable to normal human observation, the other two—electromagnetism and gravity—are not, for they are evident everywhere. Even prior to modern times, lodestones were in use as a means of allowing travelers and navigators to know true north. Since the irruption of electricity into the daily lives of humans worldwide, there seems no escaping electromagnetism in the form of digital communication, telephones, satellite television, and so on. Even the monks on Mount Athos use electrical lighting that derives its power from electrical generators in which invisible magnetic fields are constantly penetrated by wire coils, which induces an electrical current within the coils. The last of the four forces is gravity, so common as to be unobserved, taken for granted like a small plant on a kitchen windowsill. Yet its cumulative power, the gravity of each particle of matter combined into a planetary mass, exerts such force as to define the shape of the universe.

The four forces have each been experimentally observed and their particular characteristic measured, and mathematically described by means of equations that describe their behavior in controlled circumstances: gravity, first by Galileo and Newton; electromagnetism, by Faraday and Maxwell; and the nuclear forces, in the twentieth century by many experimenters and theoreticians. By now, however, it has become possible to take the mathematical expressions of each of the four forces and combine them into larger, more abstract descriptions. In Weinberg's account, the strong and weak nuclear forces were first combined into one equation, but it has almost become possible by now to combine the two nuclear forces mathematically with electromagnetism; Weinberg (along with two other physicists who were working on the problem independently) received the Nobel Prize exactly for developing such a theory.[4] Subatomic physics has developed to the point where ultimate completion in the form of elaborate mathematical expression expressed by multiple equations, confirmed by numerous experiments, is so well-established that it is called the "Standard Model" and, according to one commentator, stands as "one of the greatest intellectual triumphs of our age."

Weinberg's first book of popular science (he had written some popular-science articles before then) is *The First Three Minutes*, which describes, as far as modern physics knows about, the actual origin of the universe. Weinberg has the habit of presenting the discoveries and theories involved as if he held them only tentatively, and as if they were, so to speak, feeble guesses at cosmogony (i.e., an account of the origins of the universe). Thus, he starts this volume not with an account of Penzias's and Wilson's discovery of cosmic background radiation—which gave, in effect, final proof of the big bang theory—but with a description of a Viking myth that involves a primordial giant and a cosmic cow. Weinberg is critical, as this "is not a very satisfying picture of the origin of the universe"—as if this primitive myth had been presented in a paper at a scientific consortium. He allows how we may smile at this myth but then concedes that "the urge to trace the history of the universe back to its beginnings is irresistible."[5] Thus, the account of the Standard Model answers an unending question that bothers humans, but Weinberg plainly thinks that this account, what he calls "a modern view of the origin of the universe," is the best, really, that mankind can produce—even as he allows that it is possible that one day it may be overcome or replaced. But he does *not* mean that the Standard Model is a myth, since it is a comprehensive scientific theory.[6]

Two elements are important to note about Weinberg's presentation—one scientific, the other religious. The scientific aspect is that the underlying motive for Weinberg's writing the book when he did, in the early 1970s, was that the two major theories of the twentieth century had become conjoined: relativity theory, which deals with the vastness of outer space, the expansion of the universe, the history of galaxies, gravity, and the curvature of space-time; and quantum mechanics, which deals with the interior of the atom, including enumeration of the properties of subatomic particles and energy transfers that explain their behavior. What happens is that in explaining the earliest origins of the universe by the expansion of one nearly infinitely small and nearly infinitely dense particle, scientists must employ quantum-mechanical description of the behavior of subatomic particles. Thus, space science and atomic science met to form a new scientific version of cosmology, which thus became no longer an ignored or forbidden subject in modern science but a fruitful new field.[7]

The religious element is implied, for the use of the Viking myth replaces the more obvious ancient account, the one in Genesis that begins with the categorical statement that "in the beginning, God Lord created the heavens and the earth." Weinberg really skirts the issue, since, as he implied, it was not his job, in a book about the scientific origins of the universe, to engage in denigration or contradiction of religious accounts.[8] But putting the Viking *Eddas* on the same level as Genesis implies something else: that religious accounts in general are disposable, having been replaced by the true source of true knowledge, namely modern physics and, more exactly in Weinberg's view, particle physics. In other writings and statements, Weinberg has made this attitude clear, for he has condemned religious belief as "an insult to human dignity," even as he has defended the nation of Israel against some of its critics, putting Weinberg in the category of people of Jewish heritage who have left off belief in the Lord and the Mosaic Law but nonetheless retain their Jewish identity in a cultural sense. Also, Weinberg has defended an explicit form of physical reductionism and so, by this means, presents a coherent view of modern science in its latest manifestation as a philosophy for our time. This deserves some further comment and investigation.

Weinberg the Reductionist

As its title implies, Weinberg's second book, *Dreams of a final theory*, published in 1993, is a more personal book than his first, and somewhat less tightly organized. The style is personal, and Weinberg writes often of himself, his career, his conversations with other scientists, his expectations about scientific discovery, and his beliefs about philosophy and God. Weinberg is never strident but is nonetheless direct, for the point of the book is not to be a confessional but to put forward both information and arguments to a popular audience.[9]

The intellectual essence of his book and of his general approach to science and things in general is based on his theory of reduction. He cites the philosopher of science, Nagel, who presented, as a prime example of reduction in the physical sciences, the example of how heat theory was "reduced" to statistical mechanics (see Chapter 7). A prime example of scientific reduction is the discovery that heat is not a separate substance (called at one time "phlogiston") that, so to speak, invaded physical bodies as they were heated over a fire, but that it could best be explained as the activity of the atoms that made up the body in question. Simply put, the hotter a physical body is, the more active its atoms are; that is, they move around faster. Since the activity of a group of many atoms cannot be precisely predicted, physicists must rely on statistical averages to determine the amount of atomic activity; hence, Nagel presented the reduction of heat theory to a field termed "statistical mechanics." Weinberg cites Nagel in this regard, for Weinberg is a wholehearted reductionist, as he explains: "For me, reductionism is not a guideline for research programs, but an attitude toward nature itself. It is nothing more or less than the perception that scientific principles are the way they are because of deeper scientific principles . . . and that all these principles can be traced to one simple connected set of laws."[10] Weinberg, however, understands what may be called the existential desire to reject this view: "The reductionist worldview is chilling and impersonal. It has to be accepted as it is, not because we like it, but because that is the way the world works."[11]

Weinberg's vision is obviously related to the work he has done in his own field of particle physics, and, like the proverbial carpenter with only a hammer for a tool, everything in the world looks to him like a nail; likewise, for Weinberg, everything that is can be ultimately explained by particle physics. Scientific reductionism is rejected, however, not only by mystical theologians (Teilhard) and existential novelists (Houellebecq) but by scientists *themselves* in some instances. Weinberg cites disagreements with the historian and theorist of evolutionary biology, Ernst Mayr of Harvard University, a colleague with whom Weinberg had a longstanding but friendly disagreement, for Mayr did not believe that the principles of biology can be reduced to particle physics.[12] The opposition by biologists to a physical reductionism is somewhat common, since they tend to think that cellular activity, evolutionary processes, and especially the whole, integrated activity of an organism cannot be understood in terms of the particles and forces that are the subject matter of physics. Interestingly, Weinberg's reductive vision has been opposed by at least one physicist whose field is condensed matter physics, including the study of superconductivity.[13]

It is fair to say that reduction is more of an ideal than a general fact of science, for while it may work within the certain tightly controllable settings within particle physics, it does not necessarily work in other areas of physics itself, much less than when the reductive ideal is applied to biology or the social sciences. But Weinberg is entranced by the ideal and by the idea that some day all of science will be condensed into one beautiful, elegant set of mathematical equations, so that in the image used by another physicist, the Grand Unified Theory (GUT) or Theory of Everything (TOE) will be one day expressed on a T-shirt.

Weinberg's reductionism as a physical reductionism is, in fact, a somewhat primitive idea, not much more than the old vision of Democritus and Lucretius brought up to date by the accession of the new knowledge provided by quantum theory. What he does not seem to appreciate is that quantum mechanics has made the classical atomistic vision so complex that it can no longer carry the earlier degree of plausibility that it has had for over two thousand years. The development of the "new physics" at the time of Galileo, and put into mathematical form by Newton, did not contradict, in essence, the vision of Lucretius; rather, it strengthened it and made it more plausible. This was particularly true because Newton's vision was mathematically precise, so that given an initial set of conditions, the outcome of a physical process could be predicted by the use of the calculus. Functionally speaking, given "x" to describe a physical measurement, after a mathematical operation of multiplication or division on it, only one possible answer can result: "y" expressed as a single value. As a result, philosophers and proponents of the new physics, such as Hobbes and Laplace, envisioned, from Newton's comprehensive account of a "philosophy of nature according to mathematical principles" (a translation of the full title of Newton's famous book), a worldview that was deterministic, mathematical, implicitly materialistic, and culturally revolutionary.

The quantum-mechanical vision of atomic theory and of the universe at large is revolutionary in another way, however—in a manner that is contrary to the mechanical worldview implied by classical physics. To fully appreciate why this is, it is necessary to understand reduction not merely by the example of the transfer of the laws of heat theory into statistical mechanics; rather, it is a process of ultimate scientific discovery in which every single scientific law from any area of science is, in principle, reducible to one common set of physical laws, which, according to Weinberg, are the laws that describe the activity of subatomic particles. There is, as it were, a reductive pyramid consisting of three large blocs: the plinth or base is the physical sciences, the second large bloc consists of the biological sciences, and at the top are the social sciences. There are intermediate fields of science that hint at the unity of all science, such as biophysics and biochemistry connecting the first bloc to the second (physical science to biology), and likewise psychopharmacology and "brain science" connecting the second to the third bloc. Within the blocs, there are intermediate sciences connecting the fields within them, such as physical chemistry, molecular biology, and social psychology. This is the complete vision that serves as the subtext of Weinberg's reductive philosophy, of his "dream of a final theory."

Weinberg's dream of a complete science reducible to the laws of subatomic physics rests explicitly on particle physics understood quantum mechanically, and herein lies the problem. Obviously, it is a much different view of things if the science at the base of the pyramid is not deterministic but something radically different, and quantum mechanics certainly is. Going back to the earlier mathematical explanation of determinism in terms of "y" as a function of "x," in quantum mechanics, given a mathematical operation on "x," the result is not one discrete value of "y" but instead a range of possible values expressed as a probability. Actually, the situation is more complicated than that, for in order to determine a value for "y" given a value for "x" in quantum mechanics, the two unknowns must be evaluated together, for the measurement of "x" can never be done by itself but only in conjunction with a measurement of "y." To determine a subatomic value such as velocity or momentum or "spin" as it resides in one particle "x," it is necessary at the same time to measure the value of a complementary particle "y." Some of the mathematical complexities involved include modeling quantum operations as if they were waves and using matrix theory, tensor calculus, and irrational numbers (i.e., the square root of minus one, to which no physical application exists). Despite the incongruities that contradict common sense and the human intellect's ability to imagine, in any clear manner, how these mathematical operations apply in nature, the one consolation is that they *work*, even as they escape the control of deterministic mathematics and the limit of conceptual imagination.

The indeterminism and contrarieties of quantum mechanics, sometimes called "quantum weirdness," has not escaped the popular imagination or the musings of philosophers and religious thinkers. Weinberg is opposed to all that, mentioning, in a somewhat dismissive manner, a book, *The Tao of Physics*, written by Fritof Capra, an author who is both an adept of Eastern religions of mystical bent and a PhD in physics. The book, originally published in 1975, has seen three editions and numerous printings and promotes Capra's view that "the philosophy of mystical traditions, also known as the 'perennial philosophy,' provides the most consistent philosophical background to our modern scientific theories."[14] Capra sees in quantum physics and relativity theory a view of the universe largely compatible with Zen and the doctrine of the Tao ("the Way"), in which separation between the knower and the known is temporary or a false impression. The fact that the measurement of subatomic processes in quantum physics and of events in relativistic space-time necessarily involves consideration of the role of the observer gives color and cover to Capra's mystical speculations. Despite Weinberg's smooth dismissal, Capra's is by far the more popular an exposition of the meaning of quantum mechanics than anything Weinberg has written, meaning that in its direct influence on our culture, quantum physics—the base science at the bottom of the reductive pyramid, according to Weinberg—tells us that the physical world and the human mind are ineluctably intertwined in a mysterious and compelling way.

Hidden History of the Atom

The history of atomic theory extends from the philosophers of ancient times to the modern scientific revolution in the eighteenth century; and to the discovery, in the late nineteenth and early twentieth centuries, that the atom is complex to the current day, when subatomic physics resists all attempts at integration into a more general theory. In ancient days, the Greek philosopher Democritus and the Roman poet Lucretius first developed the atomic theory as a counter-philosophy to that of Plato and also, especially, Aristotle, for whom nature was, in effect, a compound of mind as well as matter, whose actions moved according to a designed purpose—that is, to provide for the lives of animals and mankind. The atomic theory of the ancient atomists asserted, by contrast, that nature consisted of "atoms and the void," extremely small elementary particles moving in empty space with no admixture of mind or cosmic purpose. Physical items were collections of atoms, and such processes as growth and decay—or indeed *any* change in material bodies—were explainable as changes in the arrangement of the atoms; but the atoms themselves never underwent change. The general philosophical aspect of this theory is that by understanding the ultimate elements of nature—the atoms—and their behavior, the human mind had penetrated the deepest level of reality, and that popular beliefs about the gods and life after death were discardable because they were false. Despite its inherent materialism, therefore, atomism is a reassuring philosophy, giving its adherents relief from concern about ultimate purposes, death, God, immortality, and final justice.

In the seventeenth century, Galileo revived the atomistic philosophy as a philosophical underpinning of his mechanical theory. He was mainly concerned with understanding atoms as loci of physical forces such as gravity and impulsion from cannons, and so, at this point, atomic theory for the first time became scientific in the modern sense. It was the clarity of this idea, that the behavior of atoms and all physical bodies could be explained by the application of forces mathematically and geometrically expressed, that affected Hobbes, who made it the basis of his philosophy. In Galileo's and Hobbes's vision, however, atomism did not exclude God or his divine purposes, but it did forcefully eject God and his purposes from any direct role in nature. In the eighteenth century, Newton and Boyle found further applications of atomic theory in the development, first, of a mechanical account of all bodies in the universe and second—more importantly, perhaps—in the explanation of chemical processes that were being investigated in great detail by then. Ultimately, atoms would be identified with the elements named in the Periodic Table and the characteristics of each type of atom carefully and precisely identified by means of atomic weight and atomic number. God, despite Newton's best efforts—for he was a Puritan Christian of the same mold as Milton and the Puritans who came to Massachusetts—remained at the outer edge of physical reality. God was made by some into a "watchmaker God" who, once he had set the universe in motion, needed to do no more except let his marvelously complex machine run on its own by means of a discoverable set of mathematical laws.

Then, in the late nineteenth century, experimentalists such as Thompson and Rutherford made the incredible discovery that atoms had parts—that is, atoms are not indivisible particles of the same character throughout; unlike the shot from shotgun shells or bbs for air pistols, an atom of copper was not a hard, impenetrable, small sphere of copper. Rather, it was discovered, atoms consisted of a central core (or nucleus) surrounded by shells of electrons revolving around their respective nuclei, resembling the solar system. Subsequently, it was discovered that the nuclei themselves were congeries of subatomic particles whose behavior was unpredictable by means of ordinary mathematics such as the calculus, and so, in the twentieth century, the famous "quantum theory" was developed by Planck, Einstein, Bohr, Shroedinger, Heisenberg, and others. The discovery of quantum physics in the twentieth century parallels in a rough way the discovery of irrational numbers in ancient Greece, for it disturbed the very essence of the culture itself on an intellectual level. For just as "irrational" numbers contradicted the Greek vision of perfect harmony between self and the cosmos—of a perfect rationality pervading nature and the human mind—so, also, was quantum physics the final dislocation of the mechanical view of the universe. It was one thing to do as Einstein had: that is, to combine space and time and gravity and electromagnetism, for these relativistic theories are expressed in the deterministic mathematical equations of Newton's calculus; It was impossible, however, to establish by empirical means that the same types of deterministic mathematical laws could explain the behavior of the inside of the hitherto-solid atoms.

Within the social sciences, there is a persistent and unresolved question of how scientific research affects the phenomena being studied, since most often the "phenomenon" is human behavior or human society. Scientific research into these areas really is an intrusion, and the mode of questioning or experimentation itself will often affect the kind of answers the researcher obtains from his or her study. One researcher known to the author had, as his doctoral thesis, the study of child behavior in the classroom; he described how in order to obtain relevant responses from the young children, he had had to, so to speak, become a child himself—to act like a child so as not to appear to be an adult. In general terms, this is an example of one of the solutions to this problem, the so-called participant-observer method. The issue occurs now in physical science; the question, in general terms, is "does the indeterminism at the basis of subatomic physics exist objectively within the internal activity of the atom, or is it merely a function of the human mind's intrusion into those activities?" The relevant point is that this fundamental issue of methodology applies to quantum mechanics and thus extends from the most complex area of reality, human nature and human society, to the most basic level of physical reality in the interior of atoms. The scientific concept of the atom has developed from an ancient philosophic theory to the basis of the "new physics" in the eighteenth century to quantum-physical description today. What is discernable is that the scientific concept has become inordinately complexified, even as the philosophic concept has remained roughly the same.

Using the atom as a model or analog for the individual in modern society is not unusual, since it provides an insight into the potentially tense relationship between society and self that inevitably exists. What can be further speculated is that the atom can provide another kind of model for the relation of self to society, one that is not oppositive in its implications. While atoms exist separately as the basic elements of physical reality, they also contain within themselves a tremendous force that keeps them together so they do not disintegrate—protons here, electrons there, mesons wherever. But then again, each atom contains a less powerful force that binds one atom to another in the form of molecules that make up the complex organic compounds, such as chlorophyll and sugar, which are the basis for life itself. With the actual science in view, the atom could imply a more useful analog than that of atoms floating in the void, brought together only by accident or external force. As science understands it now, the atom and the individual contain two essential forces: first, a force that binds them within themselves, keeping their separable internal parts together (thus, we say to a person fraught with anxiousness, "Pull yourself together"); second, another force that enables the atom and individual person to extend its reach beyond itself to attach to another atom or person and thus make up not a collection of atoms or a crowd of individuals, but an entity that is a unity as well as a collective, such as molecules or families and states. Here we can infer, in an analogical manner, that chemistry provides a model for societies, as a form of atomic sociology that examines how atoms combine and work with one another. We can infer that the quantum description of the atom is a form of atomic psychology, examining the deep forces within the atom that keep it together, and also examining the fearful consequences when the atoms break apart and cause radiation and nuclear explosion. This, too, has a fearful analog in the psychopathology of extreme personal disintegration—Freud and Heisenberg in the same line of work.

EPR as a Failed *Reductio ad Absurdum*

One of the classic modes of logical argument is the *reductio ad absurdum*. It is a neat rhetorical trick when used well, often seen in political debate but also more seriously when analyzing something that is proposed but whose consequence we cannot actually know. The technique is simply to take an opponent's assertion and extend it out beyond what the opponent intends, to show that the statement or proposal is unworkable or undesirable or, extended outward enough, that it contradicts the opponent's initial proposition. A scientific example is what is known as the "Einstein-Podolksy-Rosen experiment" (EPR). One of the interesting aspects in the historical development of quantum mechanics was Einstein's unyielding opposition to it; over the years, he proposed many thought experiments to try to demonstrate the absurdity of quantum mechanics interpreted as the discovery of an inherent indeterminism in subatomic reality—the so-called Copenhagen Interpretation. Einstein, with two co-workers, developed what is probably the ultimate thought experiment—EPR.[15]

In this entirely theoretical scenario, you must imagine that two particles are observed simultaneously, subatomic particles *a* and *b*. According to the laws of quantum physics, the two particles are *complementary* to each other, so that whatever happens to one must directly and instantaneously affect the other. Thus, after the two particles are observed and specific qualities observed (*a* is positive and *b* is negative, for example) and the two particles separate, then no matter how fast or how far they are traveling away from each other, if particle *a* is observed and its characteristic changed (e.g., from positive to negative), then particle *b* must also be changed in a complementary way (e.g., from negative to positive). The absurdity, Einstein said, was that, according to the laws of quantum physics, even if the two particles were traveling in opposite directions at the speed of light and were separated by light years, observing the characteristic of particle *a* would have an immediate effect on particle *b*. And that, Einstein and his colleagues knew, is simply absurd.

The EPR thought experiment has been the source of one of the more provocative problems in the philosophy of science; I remember a question about EPR being on the final examination in a seminar on the Philosophy of Physics, led by Dr. Abner Shimony, when I studied at Boston University; I also remember not answering the question very well. In any case, what happened next to EPR was that an experimental physicist decided to find a way to test the hypothesis and carefully set up an experiment fulfilling the parameters defined by the EPR paper: two electrons were observed, sent on their way, and observed a second time after they had separated (though not by light years). The result was just as Einstein and his colleagues had "predicted"; that after separation, one particle was instantaneously affected by an observation on the other. Quite unexpectedly, the EPR thought experiment has proved to be an actual, empirical phenomenon, for what had been proposed as a *reductio ad absurdum* has, in fact, become an experimental proof; if the result was and is an absurdity, it is an absurdity that exists within nature herself.

This experimental result deserves special notice, even though it is only one of many experimentally proven phenomena that constitute quantum weirdness, which includes light made up of discrete packets of wave energy, quantum jumps of electrons from one shell to another, complementarity (meaning that two particles must be studied at the same time and never one particle at a time), and so on. Consider what made Einstein and his colleagues take for granted that the results of the experiment would be absurd on their face, that the observation of one particle would instantaneously—that is, without any passage of time and across a very large span of space—affect the other. Einstein had concluded as part of his theory of relativity that the speed of light was the fastest any physical body or wave could travel, including not only light waves but gravity; yet *this* effect took place instantaneously. Also, this effect would take place no matter how far apart the particles had traveled from each other, which made the entire thing seem magical to Einstein, trespassing beyond the mechanical idea that physical motion could be explained by the image of one body coming in contact with another. The quantum-theoretical description of subatomic matter is literally unimaginable.

Despite its seeming illogicality and literal unimaginability, the EPR effect has now become so well established experimentally that it has been given a name: "entanglement." According to a recent, semi-popular book with that name, Professor Shimony has been involved in conducting some of these experiments and providing theories to explain some of the effects.[16] Entanglement has provoked speculation about instantaneous travel; in a recent television presentation, a well-dressed traveler is shown entering a pod in Paris and then immediately leaving a similar pod in New York City. To the ironic observer, however, this presentation brings to mind not the image of Scotty on *Star Trek* beaming Kirk up from the surface of a hostile planet but of the horrifying episode in the movie *The Fly*, when the first pod contains not only the experimenter but also a fly. The discordant result is that when the experimenter leaves the second pod, his body has become entangled with that of the fly.

In reality, entanglement operates only on the subatomic level, and its apparent weirdness may be the result of our not having the sensory equipment refined enough to be able to observe subatomic phenomena; in fact, our sensory apparatuses depend on those weird subatomic processes, so we should not expect to be able to explain them except indirectly—and not then expect to be able to apply them industrially to physical bodies as we experience in the human-life world.

The Supercooled Supercollider and the Mind of God

According to Stephen Hawking, "M-Theory is not a theory in the usual sense. It is a whole family of different theories each of which is a good description of observations only in some range of physical situations. . . . It is a bit like a map. . . . To faithfully map the entire earth, one has to use a collection of maps."[17] M-theory is one of the latest versions of the Grand Unified Theory, but the insistent difficulty with the dreamed-of, and hotly pursued, final theory is that it must rest on the basis of quantum mechanics. That may be its final resting place as well, for it does not now seem likely that it is possible to consistently explain quantum physics and relativity physics under one large intellectual wrapping, and so the final theory is like a taco whose filling is spilling out onto the diner's lap. The pursuit and the unlikelihood of its success appear under the two species that characterize science, experiment, and theory.

Experimentally, pursuit of the final theory involves building an underground tunnel that is as extensive but far more complicated than the tunnels that carry subways under large cities; the tunnel must be airtight, must be roughly circular, and must be kept at a temperature far below freezing, and at certain exactly measured steps, the tunnel must be magnetized with powerful electromagnets that are timed in rapid-fire sequence. The whole point of the tunnel, which is very expensive to build, is to propel subatomic particles at light speed by running them through the tunnel many times at increasingly faster speeds until they collide with a target or another particle speeding in the opposite direction. The collision yields, for a mini-micro-second, a splashy array of short-lived particles that, it is hoped, will reveal secrets to the physicists about the interior forces and particles that constitute atomic structure—and hence, physical reality in general. This tunnel, called the Supercooled Supercollider, is the latest and possibly last iteration of the first such collider (built in the 1930s at Berkeley and called a "cyclotron"), which, though cutting edge for its day, now stands like Bell's first telephone to a modern Blackberry. The cyclotron and its successors, each one able to power particles at higher speeds and energies (such as the Betatron, built at the Brookhaven National Laboratory on Long Island), have been an effective means of discovering new particles and also, as importantly, confirming or falsifying the theories put forward by physicists about the atom and its constituent parts.

The last chapter of *Dreams of a Final Theory* describes Weinberg's efforts to have a Supercooled Supercollider built in Texas, in particular by his testimony before Congress. Building the collider under Texas's soil was a project that would cost an estimated five billion dollars, but like a Defense Department boondoggle, after construction started, it became apparent that the cost would be much more. After a certain level of expenditure had been passed, therefore, a Republican-controlled Congress cut the program off, setting aside only enough additional monies to shut the effort down in an orderly fashion.[18] In his testimony, Weinberg averred how the experiments conducted with the collider would allow physicists to "know the mind of God," which was an effective way to make clear the importance of the project for future physical research, but which also may have seemed like a form of blasphemy to certain of the congressmen who were religious believers.[19] The basis for this claim was reinforced by the search for a particle—inferred from the mathematics of the Standard Model—called "the Higgs boson," also called "the God particle," for its discovery would be the empirical evidence for a finally established version of the Standard Model of subatomic physics. Alas, the Supercooled Supercollider was not to be, at least in America; in Europe, however, the collider has been built by the CERN collaborative, made up of several European nations that contributed to its cost. Appropriately, following the identification of the physics of the final theory with "knowing the mind of God," the main administrative building of CERN was inspired by the curved, up thrust form of Gothic cathedrals. In any case, discovery of the Higgs boson by means of experiments conducted on the European collider has not yet definitely happened as of this writing, results having come in slowly and partially. It appears right now that evidence for the existence of the Higgs boson is as difficult to establish as evidence for the existence of God.

Theoretically matching the Supercooled Supercollider in its extreme complexity is the theory called M-theory, the latest and most promising attempt to construct the final theory. It is not a theory that operates on one dimension, explaining a phenomena like Brownian motion by atomic activity; rather, it is a multidimensional theory, both metaphorically and actually, since, as Hawking explained, M-theory is a theory of theories, providing a very large and possibly infinite number of mathematically possible universes. Much effort by theoretical physicists has been spent in an attempt to solidify the theory and, most importantly it would seem to an outside observer, to discover experimental means to confirm the theory; however, neither of these things has taken place. At one point, when it was first put forward back during the 1980s, the development of M-Theory (or "string theory") caused much excitement with its elaborate speculations about the existence of strings as the smallest and ultimate particle at the base of the reductive pyramid, ten dimension, alternate universes, and instantaneous time travel. Not incidentally, no empirical evidence for the strings, alternate universes, or so on has been discovered. So now, like a baseball "phenom" or an ingenue about to turn forty, whose careers have not fulfilled their early promise, enthusiasm for the theory has declined.[20] God, it has been said, is a "hidden God," and the final evidence for a final theory seems as elusive. But why? Possibly, the ability of the human intellect and human engineering may have met their ultimate limits.

Final Comments on the Final Theory

It would be foolhardy for a non-physicist who no longer remembers much calculus to comment on the particulars of M-theory or the mathematical bases of strings or any proposed final theory. But from a philosophical and cultural point of view, two points can safely be made: first, any version of a final theory will have to have an epistemological element as well as a mathematical one. Conceived as a set of mathematical equations, any scientific theory must also have an interpretation—that is, a means or at least a sense of how the mathematics, whether geometrical diagrams, algebraic equations, or tensor calculus, actually apply to a matched set of physical conditions. To take the simplest case, the multiplication of two rational numbers can be used to describe the area of a rectangle, which can be applied to discover the area of a field or any plane surface. Likewise, an algebraic formula in which a measure of time in seconds is multiplied times itself (i.e., is squared) may seem a bit weird when first introduced to high-school students, but it provides a means of describing accelerated motion and the effect of gravity on a falling body. Tensor calculus applies sets of differential equations to describe the changes in a three-dimensional solid as it changes over time, a useful means of describing air flow over the wing of a jet airplane or the deformations of space-time affected by a massive body. In every example and in every case, the mathematical tool in use must have an application, and the physicist must have a rule—perhaps merely a rule of thumb—in order to know how to apply the mathematics to the physical condition.

The same must be so in the final theory, which means that the method of application must be specified, which means that, in effect, a set of instructions will have to be part of the final theory, which finally means that a philosophy of nature and of human cognition will have to be in place. After all, the final theory, which, as its proponents state, is a "theory of theories," applies to the entire physical universe and everything contained within it. The first attempt at a final theory was contained in Newton's famous *Principia* and was presented in a geometric form with axioms and derived statements and geometric-styled proofs, but it also assumed that scientific knowledge was attained by a process of induction. Thus, its main principles, such as the law of universal gravitation, were proved by Newton as the result of a collection of examples—of the earth pulling at the moon, of the moon causing tides, of the sun controlling the paths of the planets, and so on. In reality, Newton's discovery of universal gravitation, as he himself described, was the result of an intuition when he saw an apple fall from a tree and immediately associated the apple with the moon—both were pulled at by the gravity of the earth. Just as Newton's early version of a final theory implied the inductive philosophy of Francis Bacon, so also the current attempt at a final theory must have a philosophical underpinning that gives a way of relating the complex theory—far more complex than Newton's—to the various aspects of the physical universe that it attempts to explain. The underlying philosophy, it can be inferred, is that of contemporary postmodernism, in which general patterns of belief, or theories, are said to be constructs of the human mind. This philosophy is implied by the fact that the grand theory will be a theory of theories that will allow for a variety of perspectives by physicists.

Second, while the final theory may be seen as an attempt to replace religious belief in the revealed God, it cannot replace the most important aspects of religious belief. Belief in God implies that the universe is rational because it is God's creation, but that rationality is more than just a guarantee that it runs in such a manner so that the human intellect can understand and control portions of it; contrary to scientific ideals, rationality in the religious sense has a moral element, and so the universe conceived as dependent for its existence on God as Creator also has a different "feel" to it (what psychologists call "affect"). Cause and effect applies not only to the physical but to the moral aspect of lived human experience, such that vast patterns of universal cause-and-effect relationships can be seen in human history and the history of the earth, seen as mankind's home—relationships that reveal, to the discerning mind willing to experience and understand the world beyond its physical aspect, the reality of God. There is no inherent conflict between science and religion in this point of view. Thus, belief in scientific point of view to the exclusion of religious belief is, in the postmodern manner, a choice and not the result of the demands of reason.

Notes

1. Peter Gay, *The Enlightenment: An Interpretation, Volume II: The Science of Freedom*; New York, Knopf, 1969; 126.
2. See John Caiazza, "Modern Science and the Origin of Our Political Discontents," *Intercollegiate Review* (December, 1977) on which much of this section is based.
3. Steven Weinberg, *The First Three Minutes*: New York, Basic, 1993.
4. Ibid.; 143.
5. Weinberg, *Three Minutes*; 4.
6. Ibid., 10.
7. John Caiazza, *The Ethics of Cosmology*; Transaction, 2012; 4.
8. Weinberg, *Three Minutes*, 4.
9. Steven Weinberg, *Dreams of a Final Theory*; New York, Pantheon Books, 1992.
10. Ibid.; 55.
11. Ibid.; 56.
12. Ibid.; 60–62.
13. Ibid.; 58.
14. Fritof Capra, *The Tao of Physics* (3rd ed.); Boston, Shambhala, 1991; 10.
15. Louisa Gilder, *The Age of Entanglement*; New York, Knopf; 158–161.
16. Ibid., 251–254.
17. Stephen Hawking and Leonard Mladinow, *The Grand Design*; New York, Bantam, 2010; 8.
18. Weinberg, *Dreams*; 288–289.
19. Ibid.; 256–257.
20. Lee Smolin, *The Trouble with Physics*; Boston, Houghton Mifflin, 2006; 261–288, Chapter 16, "How Do You Fight Sociology?"

Chapter 9

Edward O. Wilson's Grand Unified Theory: God and Man in the Biological Universe

Evolution and Ideology

Evolution floats in our contemporary culture like one of Dawkins's memes, not merely as a strand of intellectual DNA but as a super-meme of extraordinary appetite that seeks to swallow up every other meme, as a chameleon virus in the form of intellectual AIDS. In terms of modern political theory, evolution is an *ideology*, a comprehensive account of the universe that enables the participant to make sense of all aspects of his experience and to make a commitment to a worldview he shares with others, and which provides a basis for social action by which he can identify with his friends and recognize his enemies. Evolution is an ideology, a fact that enables it to have a visible social presence, but it is an ideology that does not side with either of the usual categories of political ideologies of being either left wing or right wing. Indeed, from the aspect of the Left-right divide, evolution is an inconstant ally, slipping from one side to another depending upon whatever cultural context it acts in. Evolutionary ideology, however, does have a content of its own and an agenda. Like the AIDS virus, the evolutionary meme is a shape changer, keeping to its fundamental agenda by inserting its ideals into the DNA of other memes, including those that are religious, philosophical, and political. Evolution is the representative, in biological form, of the Enlightenment ideals of progress, secularization, and logarithmic increase of scientific knowledge and power.

Since it has become established as an ideology, the theory of evolution need no longer withstand scientific criticism to keep its cultural authority. Thus, while its "orthodox" proponents defend the ideal that evolution is an explanation of the origin of species by natural selection, natural selection itself as a scientific explanation has become less defensible as the sole or major cause of evolutionary development, for there is a real confusion among biologists about the nature of natural selection and how it operates in the formation of new species. The reason is that as biology has learned more about nature on the detailed biochemical level, natural selection has become less and less able theoretically to explain how species actually originate. One important aspect of this failure is the controversy within evolutionary biology about what is termed "the target of selection." As biology studies living things, three ontological levels are clearly discernible, first of which is obviously the individual organism itself. But to be able to identify a single organism, biology must classify it by putting the individual organism into a general category, the category of *species*, which includes all similar organisms—and thus we now have two levels of being. But modern biology has discovered a third level, the biochemical level of the immortal gene, which, it is said, defines the characteristics that make up the species. So, as a result, we have three levels of being, and the question becomes: on which of these levels does natural selection act; what is being "selected"—the individual organism (as Darwin mostly thought), the abstract category of species or group, or the set of accurately transmittable genes that make up the individual and are shared with all other members of it species? The answer is that no one knows.[1]

One of the overlooked developments in our culture's perception of science since 1850 is that evolution—rather than relativity, the big bang, quantum theory, chaos theory, or computer logic—has become the prime cultural metaphor for scientific understanding. Of the literal icons of science, it appears that the face of Darwin—heavy browed, serious, and bearded—exceeds, in its cultural resonance, Einstein's compassionate, bright, and disheveled visage. The British philosopher and historian R. G. Collingwood pointed out almost fifty years ago that the modern view of nature is based not on physics but on evolutionary biology, and that the theory of matter itself had become biological and evolutionary. "We see the old theory of matter [Newton's corpuscles] . . . dissolving away and giving place to a new theory in which matter is essentially process or activity or something very much like life," a description that accurately reflects developments in subatomic physics.[2] When physicists use the phrase "the evolution of the universe," as they sometimes do, they have, in effect, acknowledged that our contemporary view of nature is biological and not physical. That understanding does not imply a philosophy of holism or teleology, but it does nonetheless testify to the enormous power of the evolutionary narrative. Evolution as a narrative constitutes an ideology that extends beyond biology and has become an organizing principle for any natural phenomenon, so that to place an event in a line of development is to assume that its origin and nature are explained by the fact that the event came about slowly and in discernible stages. In summary, what the proponents of evolutionary ideology offer is not only a biological theory but a narrative means for attaching meaning to the seeming chaos of our lives and human history.[3]

* * * * * *

The announcement of Darwin's theory of evolution in 1859—that the origin of all life forms could be explained by means of a simply defined mechanical principle, *natural selection*, operating over geologic time spans—provoked responses ranging from frantic denial to curiosity to glorification of evolution as a philosophy of life. Many thinkers and writers, including political commentators, social theorists, and philosophers, began to re-conceptualize their respective areas of expertise in evolutionary terms. The advantages of an evolutionary take on these subjects were twofold: first, a hoped-for, but not always fulfilled, increase in understanding was brought about by the evolutionary perspective; and second, the prestige of the newest theory of modern science, one that plainly had revolutionary implications, would add plausibility to whatever point of view such thinkers were advocating. The first applications of Darwin's theory of evolution to politics and society produced a lush variety of ideas, most of them assumed by their creators to be directly implied by evolution but which, in their variety, showed how indefinite the political implications of evolutionary theory were at that time.

Among political writers who struck out on an evolutionary path were Herbert Spencer, who, in his *Social Statics*, advocated what became known as "social Darwinism," which purported to show how a highly individualistic idea of human nature was implied by the theory of evolution, which, in turn, supported a political philosophy of unrestrained capitalism. As a study in contrast, Karl Marx was fascinated with evolution to the point of requesting Darwin's permission to dedicate *Das Kapital* to him. While Spencer took the struggle of the individual to be the theme of evolution for political theory, Marx took the struggle of the classes for political control to be the primary theme, and both viewed political conflict in morally neutral terms with the resultant cost in lives and social upset an acceptable cost for social progress. Other prominent writers included Walter Bagehot, whose book *Physics and Politics* presented evolutionary reflections on the nature of social change, including the elements of transmission of ideas, discussion, imitation, and process in government. In contrast to Bagehot's establishment politics stands Peter Kropotkin, a Russian nobleman and socialist reformer whose book *Mutual Aid* interestingly used examples from nature to correct, and not merely accept, Darwin's theory by emphasizing the necessity of cooperation among organisms in the development and sustenance of life. Later on, Teilhard de Chardin's *The Phenomenon of Man* presented a version of evolution that was teleological and even mystical but which also promoted a progressive view of politics. Karl Popper, whose stringent views on the limits of human knowledge are the basis of both his philosophy of science and his Libertarian political theory, produced an evolutionary theory of human knowledge.[4]

These writers took general themes and inspiration from Darwin's theory of evolution, but in the last sixty years, there have been significant advancements within evolutionary theory. The largest general advancement has come about by the integration of the science of genetics with evolution, resulting in a mathematically precise ability to explain how organisms transmit to their offspring the characteristics that enable them to survive by giving them a selective advantage in the struggle for existence. Another large step in the advancement of evolutionary theory came about by using genetic theory to explain not just the bodily characteristics of organisms but also their behavior, including "altruistic" acts (i.e., those that seem to destroy, rather than enhance, the ability of a single organism to survive). The not-uncommon occurrence of seemingly selfless acts, such as tigresses fighting to the death on behalf of their cubs or the existence of sterile castes ("drones") among certain insect populations (such as bees and wasps), is explained by the fact that such behavior enhances the chances of the organism passing along its genes to the next generation. Terms such as *gene pool, kin selection*, and *reciprocal altruism* are now in general use as a means of explaining the behavior of bees, turkeys, and, most importantly, humans. The result has been the growth of a new scientific subfield concerned with the explanation of social behavior based on biology, a trend that has been given the names *sociobiology*, *evolutionary psychology*, and *behavioral ecology*.

Sociobiology transcends being merely a "hot" and still relatively new scientific field, for it has become a whole social and intellectual movement that puts much more emphasis on the biological sources of human behavior than the cultural, a movement that has been reinforced by the recent development of the practical aspects of genetic theory including gene therapy, cloning, and the human-genome project. The recent trend in clinical psychology to rely more on behavioral conditioning and drug treatments rather than on psychoanalysis and depth psychology has also reinforced the notion that human behavior is, at base, biological. The belief that both theoretically and practically biology is the determinative influence in human behavior has direct political consequences; in general, the Right has been approbative of the advent of sociobiology, while the Left has resisted it. However, while the Left and the Right have responded differently to the advent of social biology, their respective responses have involved them in a set of corresponding dilemmas.

The Left's dilemma is that the rise of sociobiology has seriously undercut the rational case for a progressive and egalitarian social agenda, and the Left has been thereby forced to rely on cultural explanations for human behavior, since culture can be manipulated in a way that genes cannot (i.e., by education, raising public awareness, and the force of law). The situation becomes most difficult when genetic causes are used to explain the differences between groups—in particular among races and between men and women, rather than just among individuals—for here, social biology confronts the egalitarian agenda in its most essential part. The scientific plausibility of biological explanations based on genes and evolutionary urges forces the Left to use cultural criticism to de-legitimate the very categories of scientific explanation and the ideal of scientific rationality; even bona fide scientists such as Richard Lewontin and Stephen Jay Gould have done this, which shows just how threatening social biology is to the Left.[5] The sociobiological view of human nature contradicts the thesis of cultural determinism, culture presumably being a construct as opposed to what sociobiology asserts is the given reality of human nature. Human nature as portrayed by sociobiology is morally agnostic, and moral rules or ideals of public value of any sort will not evolve from it, especially left-wing ideals, which are often not based on observable nature and which must therefore be asserted in the face of political and cultural resistance. The sociobiological view of nature is unmoral and in essence apolitical, for it seems to prohibit both the ideals of Left utopianism and traditional natural law.

On the right, the dilemma is that the materialistic and reductive tendencies of sociobiology undercut the case for traditional, religious, or teleologically based values. While sociobiology supports the traditional right-wing criticism of social reforms that require a reformation of human nature, at the same time, it undercuts the rational arguments for revealed religion and natural right. The latter occurs because of the very materiality of genetic explanation by reference to large molecules such as DNA and through the putatively natural process of evolution, which has been extended from the origin of species, including man, to the behavior of organisms, including man, also. Thus, social biology can provide a scientific basis for the traditional belief that a definite human nature exists, but only at the cost of demoralizing traditional sources of value.[6]

Edward O. Wilson's Grand Unified Theory of Evolution

The status of social biology as a branch of modern science was established by the publication of the book that gave the field its best-known name and which attracted immediate and intense attention—namely, Edward O. Wilson's *Sociobiology*. First published in 1975, *Sociobiology* is so important and influential that its fits Kuhn's definition of a "paradigm text": a book that defines a scientific field in terms of a body of facts, theories, and techniques that acts holistically such that scientists who accept it see the phenomena of their field entirely within its boundaries. The book itself is large, even in the abridged edition, which Harvard University Press subsequently published in 1980; and while often technical, it is not so difficult as to be beyond the purview of laymen.[7] *Sociobiology* is organized encyclopedically, beginning with a theoretical section on the principles of social evolution expressed in the terms of population biology; a second part on "social mechanisms," dealing with such general behavioral characteristics of organisms as communication, aggression, dominance, roles, sex, and parenting; and a third part giving detailed descriptions of those species that manifest a high degree of "sociality," which live in groups, tribes, colonies, and the like. Included in the social species are, among others, jellyfish, which swim in large colonies; ants, bees, and wasps, the social insects that live in hives; fish, which swim in schools; wolves, which live in packs; primates, which live in close-knit families and tribes; and, finally, *man*. It was the last chapter—bringing to bear all the organized knowledge about animal behavior that had gone before in the book to a description of the human race—that made it influential among intelligent people beyond specialists in evolution and animal behavior and thus caused a large public reaction.

Both the book and Wilson himself became the subjects of vehement and sustained attacks from critics, mostly on the left, who see sociobiology as not just another biological determinism but one that validates the inherent inequities of established social structure by attributing them to unchangeable human nature and by covertly making the descriptive into the prescriptive.[8] Commenting on the sustained reaction of his book among academic left-wing critics, Wilson later wrote: "There were also important political implications. If human nature is mostly acquired, and no significant part of it is inherited, then it is easier to conclude, as relativists do with a passion, that different cultures must be accorded moral equivalency. . . . The hypothesis that human nature has a genetic foundation called all these assumptions into question"[9] In effect, Wilson's book caused a replay of the controversies that surrounded the original publication of Darwin's *On the Origin of Species*; only this time, the debate is transposed from a religious to a political dimension. Once again, the human race is reduced to its animal ancestry—not only in its bodily form but in its behavior, ideas, social forms, and aspirations—with no reference to political ideals or the transcendent and the moral realms. Characteristics that seemingly differentiate human beings from the animal kingdom are ruthlessly exposed, so it seems, as rationalizations, self-deceptive fig leaves to guard ourselves against the knowledge that we are nothing more than highly complex protoplasmic machines, the accidental result of eons of the ignorant struggles of unthinking nature. This time, however, the evolutionary case is strengthened by the acquisitions of a hundred years of additional knowledge gained from research and observation and a mathematically clear idea of the essential role of genes in evolutionary process. The Left fears current biology as it once feared classical economics—that biology implies there are objective scientific laws that severely limit the degree to which social structure can be changed to accommodate egalitarian prescriptions. For the same reason, observers on the Right have welcomed sociobiology, as indicated by the highly approbative passage in Paul Johnson's popular history, *Modern Times*.[10] In general, however, there has been far less approval from the Right than criticism from the Left, and no one on the Right has dealt with Wilson's evident materialism and its effects on standards of social value.

There has been a positive and nonpolitical response to Wilson's attempt to establish sociobiology as a paradigm, namely *an increased reliance on biology in the social sciences*. There are a variety of manifestations of this trend, not all of them directly attributable to Wilson's influence, including a large number of essays, articles, reviews, and books. An example is a volume entitled *Biology and the Social Sciences*, which is a collection of essays with titles like "Culture and Sociobiology," "Human Behavior and the Behavior of Animals," and "Economics from a Biological Viewpoint." (Not everyone is enthusiastic, however, as in an essay entitled, "The Social Sciences Cannot Be Unified with Biology" by Charles Frankel.)[11] Perhaps the most significant indication of the trend is the establishment of the Association for Politics and the Life Sciences, which has its own journal and which has organized sessions at both the national and some regional meetings of the American Political Science Association.

Wilson's whole vision is not limited to the effect of evolution on social behavior and is comprehensive enough to be compared with those of physicists like Weinberg and Einstein, who have proposed "grand, unified theories." Wilson's career has been one of astounding productivity and variety. He started as an expert on insect behavior then proceeded to the biology of islands; then he went on to his major work—one that has established him as an heir to Darwin—*Sociobiology*. But Wilson did not stop there, proceeding to publish an autobiography (*Naturalist*) and to promote biological diversity (*Biophilia*), which brought to popular attention the fact that many biological species are disappearing at a very quick rate. Toward the end of his career, Wilson published important works including *Consilience*, which gave a comprehensive account of all the fields of social science and the humanities, explaining how the principles of genetic science and evolutionary struggle lay at the heart of all the disparate aspects of culture, including religion. His most recent work concerns the highly controversial subject of *group selection* (see below).[12]

The scope of Wilson's ambition can be gauged from his attempt to set out a global epistemology, for his aim is no less than to fulfill the Enlightenment ideal of a universal scientific knowledge. *Consilience* was popular enough to reach a paperback version, but it has not been as influential as his other books, since it is not about biology and human behavior but about scientific knowledge. Here, Wilson explicitly follows Francis Bacon in an attempt to organize all of human knowledge under a scientific rubric—in this case, the evolutionary inheritance of the human genome. His approach is all encompassing, including art and ethics, and his tone is gentle and persuasive; but his aim is the same as it always has been: to comprehensively reduce our understanding of all human activity to the action of genes and the forces of natural selection. His tone is seductive, as if saying to artists, politicians, and humanists that his approach, once acceded to, is not painful; and that, once the furious debates surrounding their fields are reduced to social biology, a sense of forgetting and of peace will descend. But he does *not* say how the ongoing debates in his own field of evolutionary biology will be resolved, even the ones that affect our understanding of human evolution.

Visions of the unity of knowledge are not new in Western intellectual history, as Wilson acknowledges; Wilson's vision, however, is scientific (rather than, say, Aristotelian or Kantian), but here also, Wilson has predecessors including Francis Bacon. A scientific version of the unity of knowledge is necessarily reductive, expressed in the description of science as consisting of various fields and subfields, so that the scientific version of the unity of knowledge thesis becomes, in effect, a description of the relation of separate scientific fields to one another. Wilson argues that certain fields, such as sociology and psychology, are already, in effect, being folded into the field of sociobiology, but then he goes on to include nonscientific areas of human endeavor (art, ethics, etc.) into the social sciences. There is then a double reduction: of all the social sciences into sociobiology and then of all nonscientific areas into social science. Ethics can be explained by comparative anthropology and psychology, which are reducible to sociobiology, so that ethics becomes, in Wilson's words, "the ethics of the gene."

Wilson and other evolutionary theorists (Dennett) often remark that opposition to their explanatory schemas are provoked by fear of the existential truth that evolutionary biology reveals—that human life has no transcendent meaning; yet this response is hardly intellectually satisfactory. In Wilson's case, acceptance of the validity of the general schema that he proposes in *Consilience* relies on the reader's acceptance that a myriad of details that currently remain a mystery to science will be explained sufficiently, so that all the explanatory links from one scientific field to another—from chemistry to biology and from biology to social science—will eventually be completed. Indeed, the rhetoric that Wilson and others employ is relentlessly upbeat about the explanatory power of science in a reductive mode, even where the details do not bear it out. The philosopher of science Alexander Rosenberg has detailed the gaps in the links between social biology and empirical psychology, and Michael Behe, a biochemist, has written of the anti-Darwinian implications of the molecular arrangements essential for life.[13] Wilson makes an assumption also about the power of evolutionary theory in general, as when he states that many new species have been observed in the making as a result of natural selection, yet provides no examples or citations in a book otherwise fully documented with twenty-six pages of notes.

Further, a fully articulated reductive schema based on the list of sciences with physics at the base devolves into physics, and claims of holistic scientific rules that are special to fields of science above the base of physics are usually rudely discounted by reductivists such as Steven Weinberg, who base their reductive schemas on particle physics (see above, previous chapter). Which leaves the question for Wilson to answer: does he agree with Weinberg that particle physics is the base of the reductive pyramid, and if so, how does he resolve the principles of explanation special to evolutionary biology—such as natural selection and the holistic nature of cell biology—into the laws of first biochemistry and then particle physics? The devil is in the details, and unfortunately for Wilson, the details do not support the vast intellectual edifice he is attempting to build.

The Gene Pool Concept

Because the new field of sociobiology is founded on the two fundamental concepts of population biology and altruism, it will be helpful to comment on them. Darwinian evolution depends on characteristics arising among members of species that differentiate them from other members, which are helpful in the struggle for existence and which can be transmitted by the members to their offspring. Gerbils, which are native to the North African desert, have a light-colored coat, presumably because any gerbil ancestors that had darker coats were more easily spotted by predators and thus less likely to survive long enough to have offspring. "Fitness," in this case, is the relatively simple property of coat color, which has become a characteristic part of the gerbil species. Thus far, Darwinian evolution is easily understandable; what was not well understood by Darwin was the cause of somatic characteristics (such as coat color) and their stability and difference through generations. After all, if a species is not an "essence" permanently fixed in nature, but rather a happenstance suite of advantageous somatic characteristics as Darwinian theory implies, then the underlying cause of each separate characteristic becomes the critical explanatory point for the survival of the Darwinian theory.

That underlying cause was discovered by Mendel around 1870 and then rediscovered around the turn of the century. *Genes* are the biological causes of somatic characteristics, and there are clear, mathematical laws of genetic inheritance. Two points, at *least*, complicate the matter: first, the laws of genetic inheritance are probabilistic, not deterministic; and second, most physical characteristics are controlled by more than one gene (e.g., the human face, in which a mother's hairline and a father's eye color are but two components). Despite these and other complexities, individuals can be understood, in principle, simply as collections of their genes. One principal effect of the development of genetic theory is to strongly reinforce the tendency, already present from evolutionary accounts of human nature, to perceive human nature in a reductive manner. The concept of genes gives the reductive mentality something physical to grasp, which explains with seemingly mathematical precision all the physical characteristics of any living organism, including man, *and yet which are themselves reducible to complex chemical compounds.*

While genes themselves are true physical entities observable under high magnification, once identified, they can be intellectually dissociated from the organisms in which they inhere. This abstractive process means that genes can be considered separate entities in themselves; and if individuals can be understood as merely their genes, then an entire population can be understood as an abstract collection of genes with the individuals left out. This "gene pool" concept is essential to the "population biology" that underlies Wilson's sociobiology, for the study of gene pools is typically expressed in mathematical equations that describe such topics as the distribution of genetic characteristics, differential rates of birth and death, aging, reproduction, and population density. The precise definition that Wilson gives for evolution is "a change in gene frequencies within populations from generation to generation," which shows how closely contemporary evolutionary theory is tied to the gene pool concept and the abstract study of populations; for it is the transformation of populations, defined not as collections of individuals but as collections of dissociated genes, that now constitutes, for Wilson, the process of organic evolution. One odd result of biological science's current reliance on the concepts of population biology and gene pools is that genetics, which started out as the junior partner to evolutionary theory, has now become the senior partner; evolution is being reduced to a form of historical genetics.

Altruism

One of the theoretical problems faced by Darwin was that certain individual organisms act in such a way as to bring danger and even death to themselves in order to aid other members of their species. The obvious inference that Darwin made was that natural selection worked on the group level as well as on the individual level; that is, that an individual sometimes acted in such a way as to sacrifice itself on behalf of the kin group of which it was a member, in order to increase the probability of the kin group's survival. Such an inference can easily supported by observation of, among many possible examples, distraction displays, in which adult birds display themselves in front of predators in order to lead them away from their young. As a result, evolutionary writers began to refer to "kinship selection" which refers to self-sacrificial acts that benefit the kin group.

Nonetheless, the concept remained unclear until the application of genetics to evolutionary theory was complete, to create the "Modern Synthesis"; for now, the mother bird's act of self-sacrifice can be systematically interpreted in terms of genetic investment. In this interpretation, individuals who put themselves in danger in order to protect their kin are acting to ensure the survival of the genes they and their kin hold in common. In the instance of a mother endangering herself, since each of the offspring contains one-half of her genes, her seemingly altruistic act is really a canny means of ensuring the survival of one-half of her genes. However, if the mother were to act to protect a niece or nephew, she would only be acting to protect one-fourth of her genes (the proportion of genes shared), and while she would then be motivated to act altruistically, she would put herself in proportionally less danger. This interpretation relies on a mathematical cost-benefit expression in which the likelihood of an organism's acting altruistically is described as the ratio of the likelihood that an altruistic act will result in death divided by the degree of kinship (proportion of shared genes).

While this interpretation of altruism is conceptually clear and may appear to solve the difficulty that self-sacrificial acts among animals presents to Darwinian theory, it cannot cover all the numerous examples of altruistic acts performed by human beings. Thus, "kin selection" has had to be extended beyond the limits of immediate genetic advantage to include the benefits of "reciprocal altruism"—acts that benefit human beings who are not kin and among whom the proportion of shared genes is relatively low. It is important to note, however, that when altruism is extended to this level of generality, it can no longer be mathematically defined in terms of a cost-benefit ratio. However, Wilson has recently made the case that kin selection is a false theory and now asserts that group behavior can best be explained by the development of a social gene. Wilson's theses provoked tremendous controversy, however, among "orthodox" evolutionists.[14]

Genes and Culture

In *On Human Nature*, Wilson attempts to solve the riddle of "nature vs. nurture"—that is, how much of human behavior is determined by its evolutionary and genetic inheritance, and how much is the result of what human beings make up for themselves in their "life-world" (that is, culture).[15] The basic problem is what might be termed the "incommensurability" of genes and culture, since genes are physical structures observable by electron microscopes, whose effects on the organism can be discovered by empirical research and described by biological and chemical laws. Genes are laboratory entities, and the people who research them wear white lab jackets; employ petri dishes, spectrometers, and glass tubes; and apply for grants from the National Institute of Health. Culture, on the other hand, exists in the realm of ideas and social habits, and to discover the nature of its various elements require a type of discernment that is non-localized and haphazard. Philosophers, politicians, critics, novelists, poets, artists, and dramatic writers are the expositors and creators of human culture, and they do not wear white lab coats or any other particular outerwear by which they may be identified. Overcoming the non-locality of culture must be the first step, therefore, in any attempt to correlate the activity of genes with the shape of culture, and just as the stream of genetic material through evolutionary history can be localized by identification of the gene (the image is Dawkins's), so also the stream of *culture*, it is hoped, can be localized by some unit resembling a gene, impelling genetic theorists toward what Wilson called the "search for the basic unit of culture."[16] Wilson devised a term for the unit of culture, "culture-gen," but concedes that Dawkins's popular term "meme" has carried the day.

Wilson attempts to broach the split between what he terms "nurturists" and "hereditarians," for his theory is subtle in that it proposes not a direct dependence of culture on genes but, as he terms it, a process of "co-evolution" or gene/culture interactionism, which cedes to culture effects on behavior that influence the evolutionary development of human beings. But in the end, the genes are primary for Wilson, and his materialism is relentless as he concedes no possibility of an essential reality of culture. The correlate to a gene or the action of a group of genes must be, in effect, a discrete particle of culture for this approach to succeed, but the elements of culture are usually understood to be ideas, which are not easily analyzed into separate constituents. Terms like *culture-gens* and *memes* are, at best, analogies or metaphors. The efforts of someone like Daniel Dennett, the philosopher who has written a virtual hymn to Darwin's theory of evolution, to describe the essence of memes as existing in some indefinite "space" indicate the difficulty of precisely identifying what they are.

Wilson carefully distinguishes the degree of genetic determination that applies to the different types of human behavior, varying it from topic to topic as the evidence allows. Wilson goes so far as to provide charts that show the degree to which characteristics are either unique to the human species or are shared with primates and vertebrates.[17] It is important to note that for Wilson, genetic indetermination of a particular characteristic does not mean that it is random and thereby beyond the scope of evolutionary process; such indeterminateness only means that the evolutionary process is transposed to another level, albeit one in which more variability is apparent. Thus, he refers to culture as an evolutionary tracking device and gives an overly brief but entirely reductive amount of religion, explaining it in terms of individual and social survival.[18] It is also important to note that Wilson relies on cultural anthropology to assess the degree of genetic influence that applies to a given human characteristic (i.e., how much of a human trait comes from animal ancestry). But the assessments of cultural anthropology are based on the observation and comparison of gross cultural characteristics, not on the chemical components of specific genes, and are far less likely to be as accurate, since such observations and comparisons will be colored by preexisting beliefs held by the observer—a point that has not been missed by Wilson's left-wing critics.

Sex and Religion

Human behavior is driven by many types of motives, from the most inherently biological to the most transcendently human. In *On Human Nature*, Wilson considers the two most representative factors relevant to the understanding of human society: sex and religion. Wilson begins his account of human sexuality by denying what most people take for granted—that the primary functions of sex are reproduction and pleasure. The reason for the evolution of sex, Wilson says, is to create genetic diversity, for a simple division of the organism would result in an offspring that has the exact same genetic, and hence physical, characteristics (i.e., a clone). Wilson, apparently anticipating resistance among his readers, devotes a good deal of space detailing the degree to which sexuality is biologically ordered. The point is especially important, because the variation in sexual mores and in the division of labor between sexes among the thousands of human cultures is enormous. However, even though the variation "is based on culture . . . [n]evertheless, this flexibility is not endless."[19] Wilson summarizes this evidence in a minimalist manner, characterizing the biological determination of sex roles as a twig "bent a little bit," as "not entirely an accident of cultural evolution," and as a "slight biological component."[20] However, the biological determination of gender roles is a fact, which leaves only limited options for public policy, since eliminating all sexual differences is likely to involve prohibitive social costs. Wilson obviously anticipates a feminist rejection of this, calling the genetic determination of sex roles, "one of the most inconvenient and senseless" relics of our prior genetic history.[21]

Wilson is on the horns of a dilemma here, for the biological view of human nature results in what may be called "descriptive standards"; that is, the biological imperative gives us a destiny that we must face if we are to respond truthfully to the facts of our nature. They are not exactly ethical norms in the sense that they are not externally given (as, for example, the Ten Commandments are), nor do they have a moral obligation attached to them. Yet such descriptive standards imply that how our nature is constituted leads us to act in a certain way that we can only contravene, as Wilson points out in the instance of gender roles, at a very high social and individual cost. Thus, we are not yet to the point of a traditional ethics of sexuality based on "natural law," but we are close enough to it to make Wilson cautious about what he describes as the biological function of sexuality.

Wilson's stringent criticism of traditional Jewish and Christian rules for sexual behavior is him "protesting too much." For he can evade giving traditional sexual morality credence from his evolutionary point of view only by ignoring the two most obvious facts about human sexuality—as the means of human reproduction and of the emotional bonding of couples. If sex is not, in essence, a means of reproduction, it is difficult to say *what* it is; what else do the complicated organs of sex exist for but the destined result of copulatory sexual activity, namely offspring? Why deny in this context what he asserts so vigorously in every other context—that human behavior is determined by the need to transmit genes? As for bonding, Wilson waxes almost lyrical about how the lack of pronounced estrus among human females allows for nearly constant sexual contact and its effect on cementing long-term family bonds among humans. But to state that the biological ends of sexuality are reproduction and bonding is to support the validity of traditional sexual ethics; and for Wilson, this was obviously too close for comfort.

Religion is linked with sex, typically surrounding the subject with taboos, rules, and commandments; but beyond constraints, religious doctrines often regard human sexuality as connected in some way with the divine. Regarding religion, Wilson states that he faces his greatest challenge in his attempt to construct a grand unified theory extending from biology to human society. Not only is religion "the most complex and powerful force in the human mind," it is also far removed from the main principles of sociobiological explanation, based, as they are, on "population biology and experimental studies in lower animals." Further, religious motivation is probably "hidden from the conscious mind" because religion is, above all, "the process by which individuals are persuaded to subordinate their immediate self-interest to the interests of the group." Despite the persistence of religious belief in the face of modern scientific progress, its seeming paradoxes can eventually be explained "if we pay due attention to the sociobiology of religion."[22]

The chief effect of religion is the willing subordination of the self to the group, which has benefits for *both* the individual and the group. There are tensions implicit in the practice of religion, since religious codes, which are often very strict, operate at the level of group selection (promoting the survival of the group)—yet Darwinian selection operates on the level of individual selection. However, group and individual selection need not be mutually exclusive, for the average individual gambles that subordination to the religiously defined group will ultimately benefit him. The principle advantage of religious belief for the individual is the strengthening of his own sense of identity. This "sacralization of identity" predisposes the mind—"one can speculate that the learning rules are physiologically programmed"—to a few steps, "which in combination generate the institution of organized religion." Wilson states that "human beings are still largely ruled by myth," and among the disposable myths is the God of monotheistic religion—who is always male, and who is always found among the beliefs of pastoral societies where herding and the figure of the shepherd are the primary features. Another disposable myth is Marxism, which masquerades as a form of scientific materialism but is not, since it rests on untested premises, principally the notion that human behavior is unstructured and can be molded by social environment. "Marxism is sociobiology without the biology." As Wilson evaluates them, Marxism is passé, and religion is still strong but can be explained by scientific materialism, which is the myth that will prevail, even though "it suggests only an existential meaning for the human species."[23]

Wilson's treatment of religion is perhaps the most unsatisfactory portion of his sociobiological edifice. Since he admits that his view of scientific materialism stands or falls on the success of its reductive treatment of religion, Wilson's failure here indicates the failure of his entire effort. The chapter on religion lacks coherence, consisting mainly of a succession of topics incompletely explained and not always related to each other. Wilson understands his dilemma well—that religion is so far removed from the main principles of sociobiological explanation that, in order to account for religious belief in an evolutionary manner, he must extend his use of "hypertrophy" and the unconscious well beyond any previous limit. The link between religiously inspired behavior and improvement of the gene pool is so tenuous that Wilson invents whole new categories of selective process to account for it (e.g., "ecclesiastical selection"). He cannot provide a consistent explanation of the selective advantage of religion, because it reaches extremes of both group and individual selection, producing a conflict that Wilson never overcomes.

Mind and Purpose

In *On Human Nature*, there is a passage in which Wilson is explaining the process of hearing in a human infant, the infant being described in a subsequent passage as "this marvelous robot." Wilson traces the sound from the point where it begins as waves of compression in the atmosphere to "receptors in the inner ear" to electrical signals passed to "corresponding masses of nerve cells at successively higher levels of the brain"; to way-stations in the hind-brain, to inferior colliculi of the middle brain and medial geniculate bodies of the forebrain, and finally, to the auditory cortex, "where in some manner beyond our present understanding the mind 'hears' the sound."[24] It is a revealing passage—an exact scientific description of the process of hearing followed by Wilson's frank admissions that he really doesn't know how to explain the last step, which is the key to the phenomena, just at that point where the child's mind comprehends the event by turning sound waves into meaning.

Most surprising in Wilson's treatment of the human species is the total exclusion of mind or purpose to explain human social behavior. This, however, is a reflection of an exclusion that characterizes his treatment of all animal social behavior, a neglect criticized by geneticist C. H. Waddington in his review of *Sociobiology* as "the weakest feature in the whole grand structure which Wilson has built up . . . [I]s it not surprising," Waddington asks, "that in a book of 700 large pages about social behavior, there is no explicit mention whatever of mentality?" After quoting a passage from *Sociobiology* that describes the distraction display of a certain bird species that fakes an elaborate death to draw predators away from its nest, Waddington asks further, "[c]an there be any point in renouncing the use of words like mind, or goal-seeking, in the face of phenomena of this kind?"[25]

If the exclusion of mind amounts to a neglect in Wilson's account of animal behavior, it amounts to distortion in his account of human behavior. One result is that those human characteristics that Wilson says are "unique" (such as communication and role playing) cannot be adequately explained sociobiologically, precisely because the element that makes them unique—mankind's intelligence by which to understand symbols and project purpose—is denied. Among the examples of human uniqueness, however, one in particular stands out, for Wilson must account for the later evolution of the human race, the transformation of mankind from a being barely distinguishable from an upright ape to the creator of advanced civilization in just ten thousand years, an extremely short span on the evolutionary time scale. As Wilson puts the problem, "[b]iological parameters usually change in a rate-dependent manner to slow growth and eventually bring it to a halt. But almost miraculously, this has not yet happened in human evolution." What might explain later human social development, which, since AD 1400 in Europe, has expanded "not just exponentially, but super-exponentially"? Wilson cites five causes, including sexual selection, multiplier effects in cultural innovation, agriculture combined with increased population density, warfare, and "multifactorial systems" (i.e., the joint effect of the foregoing four causes).[26]

While Wilson assumes that intelligence has continued to evolve during this process, he nowhere cites human intelligence or the development of mental capacity, as indicated by the exceptional size of the human brain, as a cause. Indeed, he seems to deny it when he states that there is "no reason to believe that during this final spurt [of human social development] there has been a cessation in the evolution of either mental capacity or the predilection toward special social behaviors . . . it would be false to assume that modern civilizations have been built entirely on capital accumulated during the long haul of the Pleistocene." In other words, intelligence and the mind are an effect of social evolution—gradually increasing along with it—rather than a cause, even though Wilson admits, "we do not know how much mental evolution has actually occurred."[27] Wilson does not realize that "mental evolution" at such a rate is no more inexplicable than is the social evolution that he supposes caused it and for which he only suggests possible causes, the intellect not among them. But why not posit mental evolution as, at least, a *possible* cause of what Wilson admits are inexplicably unique social characteristics among humans, such as communications and roles? The only answer can be a prejudice on Wilson's part against mentality, intelligence, and mind, as such.

Wilson must find substitutes for the mind as purposive agent. Darwin used "natural selection," but Wilson's choice is hypertrophy, "the extreme growth of pre-existing structures" from the animal to the human spheres, supplemented by a theory of unconscious motivations.[28] In effect, human beings act the way they do, even at their most enlightened and self-conscious, because they are driven to by the intensification of animal urges, which they do not recognize, because these urges operate on the unconscious level. Wilson does not explain why these urges, so enormous in their import and effect, remain unconscious, nor why characteristics pre-existent in animal behavior become more intense among humans—so that, for example, bonding becomes religion, and altruism becomes sanctity. At this point, it is fair to say that the suppression of mind as an integral aspect of human life and social evolution results in a gross caricature. In Wilson's portrayal of human nature, human beings are given no credit for intending their own actions or for having made choices at particular moments in history that have influenced the character of their culture.

What is it about the *mind* that so puts off Wilson—and not only *Wilson* but also other social evolutionists, including Dawkins and Dennett, and most other social scientists, including psychologists, anthropologists, and sociologists? The answer seems to lie in the nature of scientific inquiry itself, in its methodology and its basic premises. First, the introduction of mind (or "will," "intention," "purpose," etc.—all the words that designate that the human individual acts as an agent on his or her own) is taken to mean that in the context of social science, an inherently unpredictable element has been added, a knot or a surd irresolvable into scientific laws, which therefore will invalidate any attempt at scientific generalization about human behavior. ("Why should human behavior be any different from any other phenomena?" the social scientists ask.) Second, if the mind is a real element in human behavior, it stands as one of those inferred entities, like a gene or an atom, which are beyond the scope of our senses but that still have real existence. However, in the case of mind, unlike genes or atoms, the nature of the inferred reality is immaterial and not such as science deals with, but is rather "meta-" or "trans-scientific" in nature. Wilson, other social biologists, and most social scientists deny, as a matter of principle, the existence of immaterial mind as an agent of human behavior; but such a principle is, at best, a working assumption—and, at worst, a philosophic prejudice disguised as unprejudiced science. Science can only presume or hope that all of the phenomena it studies can be described by its methodology; it is not *necessary* that it be so. Further, the existence of mind does not prohibit generalizations about human nature; however, these generalizations will not be scientific in nature, but moral. "Ought" will not always be separated from "is," no matter how disconcerting Wilson and others find that to be.

The great significance of sociobiology, according to Wilson, is that it connects social science to biology and, for the first time, gives a truly empiric ground to social science. Further, while Wilson writes of *logotaxy*, the acquisition of knowledge over time, science is not among those aspects of human existence, such as aggression or sexuality, that Wilson tries to explain.[29] But why not? A sociobiological explanation of modern science can easily be imagined; anthropology provides us with examples of primitive peoples supplying myths to explain the origins of the universe (as Weinberg noted); there are significant indications in animal behavior of primate investigations into the animal's physical environment, not all of them presumably done as a search for food but in a disinterested way—investigation for its own sake, so to speak.

Why isn't intellectual curiosity among the general aspects of human nature that Wilson, social scientists, and evolutionary biologists prefer to study? For Wilson, presumably, the contrast between the scientist striving disinterestedly and methodically after pure knowledge versus woodpeckers hammering on bark to dislodge insects for food, or monkeys placing boxes on one another to reach a banana suspended from a ceiling, would not support the proposition that science is a hypertrophic extension of some form of animal behavior. But then, this is true of other types of human activity as well, such as religious worship, which Wilson says is the most intractably human of activities. For if Wilson were to treat science as one of the chief characteristics of human history as he does other human activities (i.e., reductively), it would invalidate sociobiology itself along with all other kinds of knowledge. Indeed, Wilson claims that evolution is no more than a myth, albeit the myth most likely to reach whatever truth the human mind is capable of reaching. But then, having reduced knowledge to a biological phenomenon, how can Wilson be said to *know* this? The denial of mind undermines Wilson's grand unified vision.

Material Mankind in the Evolutionary Universe

"The core of scientific materialism is the evolutionary epic," says Wilson, but it is also a myth whose most sweeping assertions cannot therefore be proved with finality. It remains, however, the best myth we are ever likely to have, since it is "as close to the truth as the human mind is constructed to judge the truth." Wilson contends that science will overcome religion and is annoyed that the high culture of Western civilization ignores sociobiology, writers and thinkers not realizing its importance and value.[30]

Scientific materialism is not new, and the objections to it are well known; the implicit hope of Wilson's unified theory is that the recent discoveries of sociobiology would be enough to overcome them, but they are not. Wilson does acknowledge the inevitable dilemma in his position: that scientific materialism cannot provide a transcendent purpose for mankind to pursue or by which to judge its actions or control its desires, a ramification that leads him into a vulgar existentialism, for quotes from Camus both begin and end *Sociobiology*. Stephen Gould, in an article on the sociobiology of altruism entitled "So Cleverly Kind an Animal," approvingly cites these words of Simon de Bouvier: "The essence of man is that he is the being that has no essence," the poverty of modern thought thus expressed in the contradictory vocabulary of classical metaphysics. That existentialism and the latest developments of biological science have the same implication for our view of the human person may seem odd, yet the cause and the effect are the same. The cause is the denial of known purpose as a part of human life, either because of a sense of failure in European culture (in the case of existential philosophy) or because transcendence defies empirical norms (in the case of biological science).

The effect, in either case, is the denial of the reality of the human person, most explicitly drawn by Wilson when he states that the individual is evanescent—that only genes, in effect, are real. For Wilson, the gene pool is the one surviving constant of evolutionary process, which he has extended from its proper place as an abstraction of evolutionary science to a basis for norms of human behavior. But these norms are not really moral norms, lacking the aspect of moral compulsion that prescriptive moral rules have, which leads to another thing that Wilson shares with the existentialists—that is, the reliance on human will to break out of the circularity of the dilemma posed by the denial of purpose and the need for action. With the denial of the ability of the human person to know, to conceive, and to carry out an effective plan of action comes also the somewhat-contradictory reliance on pure acts of will, unconnected to any understood human purpose.

The question of social values is critical for Wilson. The two essential questions are: whether there are universal standards of human behavior; and, if so, what their nature and origin are. Wilson answers that there *are* universal standards of behavior, and that their nature is based on their evolutionary origin. Although systems of human value share universal characteristics, as dealt with by Wilson, they are not normative but descriptive: the universal human tendency to think that standards of value arise from an "external" or transcendent source is, in effect, the rationalization of a biological urge. He dismisses the possibility of a source of values external to biology out of hand, expending considerable intellectual effort providing reductive explanations of the origins and operations or value systems—as is evident in his treatments of altruism, religion, and sexuality (see above). It should be emphasized, however, that Wilson's evolutionism provides a consistent explanation for the manifest persistence of belief in universal systems of human values, in contrast to most social scientists, who accept and often promote a view of cultural relativism. The reason for the difference is that Wilson's evolutionary view of human nature, while materialistic, ascribes to it definite characteristics, seeing human beings as having a vast inheritance from their animal ancestry and exercising their wits and energy in social labor while fulfilling themselves individually according to biological lineaments.

Wilson has established that there is a biological basis for human nature, which is enough in itself to have enraged his left-wing critics, because nature implies a set of definite characteristics and limits to the imaginable arrangements of human nature and society. However, that nature, as Wilson describes it, is mechanical and totally biological; it is human nature, with more emphasis on the "nature" than on the "human." In Wilson's view, a woman need not be a mother to fulfill herself biologically, but only because spinsterhood fills a biological need that is of adaptive value to her kin. Yet even a reductively inclined description of the biological nature of the human race implies that some types of behavior are more beneficial for it than others, leading to some kind of norms. However, the norms can hardly be called moral principles, which raises the question: how close is Wilson to a position resembling natural-law philosophy? The old biology on which the medieval principle of natural law was based was teleological, for its descriptions of natural phenomena took in not only material and efficient causes but formal and final ones as well. But Wilson's biology is relentlessly mechanical, suppressing or neglecting any evidence of nonmechanical reality, even where it is most apparent and even when it leaves his account of human nature more a caricature than a portrait. From this point of attack, traditional natural law cannot evolve, the sundering of fact from value accomplished by the construction of a Berlin Wall of scientistic prejudice.

But Wilson's caution when describing human sexuality brings out something characteristic about the evolutionary point of view—namely, how difficult it is to maintain the separation between the fact and value in biological descriptions of human nature. The word "nature" tends to take on a purposive connotation, while the expression "law of nature" begins to bear a moral as well as a scientific meaning. Of course, as a committed evolutionist, Wilson would deny that, yet his awareness of the proximity of a teleological interpretation is evident, for instance, in his attempt to find a selective advantage in homosexuality.

Throughout his entire discussion of human evolution, Wilson has made the case for the development of the human race in such a way as to avoid any suggestion that it was purposive. To do this, however, he has had to deny evidence of intentional activity on the part of human beings, either individually or socially, and to evade the consequences of his own description of human biology. This tension between the facts that imply purpose and Wilson's prior commitment to a mechanical view of human life results—in *On Human Nature*—in a style that is often discursive and inexact, occasionally even relying on bluster and emphatic denial. This is not the case in *Sociobiology*, where Wilson is well organized and always exact, and where he almost cheerfully acknowledges the limits of present sociobiological explanation when applied to human beings. But in *Sociobiology*, he is addressing his fellow scientists, while in his later, popular works he is completing his grand vision of a sociobiological universe and addressing the general public, who apparently should not be made aware of the significant weaknesses in the case for the mechanical view of human evolution. But then, in his later works, he is writing speculatively—as a materialist philosopher and not as a scientist.

Notes

1. See Edward O. Wilson, *Sociobiology, The Abridged Edition*; Cambridge, Harvard U. Press, 1980; 50, where the issue is presented in terms of "group selection." See also George C. Williams, *Adapatation and Natural Selection*; Princeton, Princeton U. Press, 1966, which is the canonical start of discussion of the issue of group versus individual selection.
2. Collingwood, Robin George, *The Idea of Nature*. Oxford, Oxford U. Press, 1960; 132, 147.
3. John Caiazza, *The Ethics of Cosmology*; New Brunswick, Transaction, 2012; 35–48.
4. See John Caiazza, "Political Dilemmas of Social Biology," *Political Science Reviewer*; Vol. XXXIV, 2005.
5. *The Sociobiology Debate*, ed. By Arthur Caplan; Harper and Row, New York, 1978; see among others, Elizabeth Allen et al. (the "Sociobiology Study Group)," 280–290, "Sociobiology: Another Biological Determinism;" 280–290. For a methodological left-wing attack see *Not In Our Genes* by R.C. Lewontin, Steven Rose and Leon Kamin; Pantheon, New York, 1984.

6. For conservative reaction to sociobiology see *The Politics of Human Nature* by Thomas Fleming; New Brunswick, Transaction Books, 1988. Also see *The Conservative Movement* by Paul Gottfried and Thomas Fleming; Twayne Publishers, Boston, 1988 which recommends an attempt to utilize sociobiology to help establish a theoretical basis for conservatism.
7. Wilson, *Sociobiology, The Abridged Edition.* This edition is referred to throughout.
8. A rally against Wilson and his ideas was held, excoriating Wilson and his ideas as characterized in a handbill, distributed by "ICAR: the International Committee Against Racism," in author's possession.
9. Edward O. Wilson, *Naturalist*; Washington, D.C., Island Press, 1994; 334.
10. Paul Johnson, *Modern Times*: Harper Colophon, New York, 1983; 732–734.
11. Thomas W. Wiegele (ed.), *Biology and the Social Sciences*; Boulder, CO; Westview Press, 1982.
12. Edward O. Wilson, *The Social Conquest of Earth*; New York, Liveright, 2012.
13. Alexander Rosenberg, *Instrumental Biology, or The Disunity of Science*; Chicago, U. of Chicago Press, 1994. Michael Behe, *Darwin's Black Box*; New York, Free Press, 1996.
14. A general account is in the *Atlantic*, Howard W. French, "E. O. Wilson's Theory of Everything;" http://www.theatlantic.com/magazine/archive/2011/. . . wilsons. . ./308686. A critical account is contained in Steven Pinker's article "The False Allure of Group Selection" in the on-line journal *Edge*; http:// edgeorg/conversation/the-false-allure-of-group-selection (accessed 5-26-2012).
15. Edward O. Wilson, *On Human Nature*; Cambridge, Harvard U. Press, 1978.
16. Wilson, *Consilience*; New York, Vintage, 1998; 146.
17. Wilson, *Human Nature*, 276, 291.
18. Ibid.; 284–287.
19. Ibid.; 125.
20. Ibid.; 132.
21. Ibid,; 135.
22. Ibid.; quotes found on 169–172.
23. Ibid,; quotes found on 188–193.
24. Ibid.; 54.
25. "Mindless Societies" in Caplan, *Sociobiology Debate*; 252–258.
26. Wilson, *Human Nature*; quotes found on 296–299.
27. Ibid.; 296.
28. Ibid.; 89.
29. Ibid.; 169.
30. Ibid.; 201, 203.

Chapter 10

Stanley Milgram's Famous Experiments and the Awful Authority of Social Science

On Social Science in Contemporary Culture

Despite the intellectual appeal of physical reductionism and the practical influence of genetic science, the contemporary age is primarily the age of the social sciences. Just as in Newton's day, when mechanical explanations of human behavior and cosmic order were heard in the salons; just as in Darwin's day, when evolution was heard in the high councils of European governments to justify colonial expansionism; so in our own day, the techniques of behavioral psychology are used to regulate the behavior of the general population. We speak knowingly of Freud's theory of sexual repression and of the social psychology of interpersonal relations; our courts summon psychologists to determine the fitness of suspects to be tried for felonies; and our politicians seek the expert advice of sociologists, economists, and political scientists on policies that affect the lives of millions. Our art reflects themes drawn from social sciences—Munch's *The Scream*, a portrait of ultimate despair, or Giacometti's sculptured figures, denuded of humanity. Withdrawal, alienation, and social maladjustment replace tragedy and the conflict between good and evil as the subject of our plays, novels, and films; *Krapp's Last Tape* replaces *Macbeth*, Gregor Samsa replaces Ulysses, *The Wasteland* replaces *Paradise Lost*. What spell have we got ourselves under? What is the internal nature of this social science that seems so well to reflect our anxieties, needs, and purposes in the late twentieth and early twenty-first centuries—this psychology of psychology, so to speak?

The advocates of a social-scientific view of mankind and human society presume that the empirical method of modern science is the only valid means of understanding human psychology, history, society, economics, politics, and personal relations. Further, they identify the proper use of reason itself with scientific method. In effect, they have developed a political and social philosophy in which rationality is exclusively their own and defined in such a way that its expression/exercise always implies the validity of their prescription for social ills. Debate with other schools or other approaches is excluded by this mindset, since the advocates of behaviorism presume—and do not even try to prove to themselves by a conscious argument, but already *know*—that they are in control of all the processes of social thought. This is an insidious mechanism of self-justification that means that debate with social-science advocates is not on the same order of debate between, say, management and labor, or even between social conservatives with progressive reformers. Rather, debate can become the moral equivalent of war, in which it is treason to even listen to the blandishments of the other side.

That the parallel is not overdrawn can be seen in the response of social-science advocates to those who have opposing cultural and political views; for rather than develop arguments against, for example, advocates of a religious or moralistic point of view, these opponents are characterized according to a psychosocial typology that renders argumentation superfluous—for you do not argue with someone under the influence of psychological malady or an embittered social conscience. The foremost example of such categorization is the famous "Authoritarian Personality" that was originally constructed by the Frankfort school of social psychology to explain why people followed the leadership of Hitler and the Nazis, but which is now more often used to derogate conservative ideological opponents. Another example is the use of the concept of "status envy" to explain middle-class opposition to social reforms.

The two most influential theories based on what is now often recognized as a misuse of social science are those of Karl Marx and Sigmund Freud, both of which exhibit this effect. Marxists typically short-circuit debate on issues of revolutionary violence or social justice by accusing their opponents of arguing from self-interest as members of the bourgeoisie; after all, defenders of the status quo inevitably are its beneficiaries, and so, for instance, statistical analyses of the cost of welfare programs to overburdened tax payers is claimed to be the result of identification with oil millionaires. Freudians use the same general technique; however the categories are different, as in "unresolved Oedipal complex" or "obsessive neurosis." Family conflicts result in the overuse of psychological explanations, as when siblings in conflict claim "Mom always liked you best." A boy's best friend may be his mother in the old expression, but nobody uses it now since the advent of Freudian influence, which is as much a social pathology itself as the psychological pathologies it attempts to resolve. Importantly, both Marx and Freud were atheists, which leads to the earned suspicion, upon considering the extraordinary comprehensiveness of their *faux* scientific accounts, that they have replaced the comprehension usually reflected in religious and grand philosophic narrative with their own theories—pretension and hubris, it may be said, in that they attempted to replace either the Holy Spirit or the demiurge of history with their respective accounts of class warfare leading to social justice and of resolution of repressed emotional conflicts leading to social amity.

But more common and relevant are the social-psychological and recent evolutionary explanations of religious belief. Here, there is a panoply of current explanations, all of which aim at de-legitimizing religious faith by connecting it with various nonspiritual causes—psychological instability; inability to cope with the turmoils of present-day life; the desire to simplify the understanding of reality and human behavior by employing dualistic categories of absolute good and evil; and, most recently, as the result of a genetic predisposition, as in the assertion that there is "God gene" (or, more subtly, as the result of a complex of evolutionary causes that have accidentally left religious belief, like an appendix of the psyche, as a nonessential but still-existent organ).

The problems accelerate in the real world of social policy, beyond that of mere debate, when clashes among various segments of the culture over social issues frequently involve how much power each will have in forcing its value system upon the general population—and, more pragmatically, how much government favor in the form of grant money these segments will have. Indeed, cultural conflicts frequently concern the status of values; sexual morality is the obvious instance, for it is a dominating element in debate about, for example, sex education in the schools or the general availability of birth control. The social-scientific view that an enlightened attitude, guilt-free experimentation, and knowledge of the facts of reproduction are the keys to sexual integrity contrasts sharply with the traditional and moralistic view concerning nice girls, virginity, and the proper role of sex in family life.

The final effect of identifying scientific method with rationality and goodness is that those who advocate social policies on this basis have the peculiar attitude of people who are sure that they have got hold of reality when they really haven't. That is, they tend to be dogmatic, intrusive, and arrogant. This group of advocates, singularly unprepared to cope with challenges, must meet them nonetheless, until the whole world outside of their protective environment of social-scientific ideas becomes for them a place where, as they experience it, they meet a constant succession of unprovoked attacks. Such "good thinkers" become suspicious, imputing conspiracies by right-wingers, Christian fundamentalists, and military leaders until authority itself becomes a perceived hazard to them. As is often the case with suspicious people, however, the opposition that is generated against them is a reaction to their own actions, for no one would have cause to vigorously oppose advocates of social science if they were a philosophic school and nothing else. It is in the application of policy science, econometric models, and sociologies of political typologies that has led to the direct involvement of the social sciences into every aspect of human life—and hence, of opposition to them.

From their early origins, the social sciences have shown a concern for the control, as well as the understanding, of human affairs. Bacon said that knowledge is power, expressing in that aphorism one of the two underlying purposes of the growth of all of the empirical sciences (the other is understanding for its own sake); and what is true for the physical and biological sciences is true for the social sciences, as well. In this sense, the purpose of the development of the social sciences is political, a fact that can be seen in the outlook of those theorists who created the concept of the social sciences, such as Comte, Sainte-Simon, and Hobbes. The purpose of control of the social environment continues, stronger than ever; only now, as seen in the uses of technology and propaganda used by modern totalitarian governments, the concept of social control—which was at first vague and undefined by empiric references—has put in the hands of governments the power to control, as well as to explain. In a contrary manner, it was the famous experiments on the nature of authority by Stanley Milgram that illustrated what was the means, but not the point, of his experiments: that is, the authority of social science, which I have termed the "awful," in the double sense of being both fearful and awe inspiring.

Stanley Milgram's Experimental View of Obedience to Authority

In the early 1960s, psychologist Stanley Milgram, then of Yale University, conducted a series of experiments on the behavior of human beings, the purpose of which was to examine the phenomenon of obedience to authority. The results were startling, showing that human beings would inflict needless suffering on innocent fellow beings simply because an apparently legitimate authority commanded them to do so. According to Milgram and others, these results explained how it was possible for Hitler to command the obedience of men to torture, kill, and destroy millions, and they also showed the necessity for citizens to be far more careful about obeying the commands of their governments—and those of any other authorities or authority figures, including, presumably, parents, schools, police, and clergy. Milgram's argument therefore utilized the results of scientific experiment and scientific methodology to prove an explicitly political point about how human beings ought to regard authority. Ironically, what is left out of Milgram's conclusions from these experiments is that they depend upon and reinforce the authority of science in contemporary culture, especially social science.

Milgram's experiments on obedience are possibly the best known of all those that have ever been carried out on the behavior of human beings. They have a shocking quality, as if a veil has been ripped aside, leaving us to stare at the stark nakedness of human nature, revealing details that horrify and embarrass us. The interest that they aroused in the world of social science is revealed by the number of initial responses written by social scientists at the time and the number of references to Milgram in the literature up to the present. There have been several books triggered by Milgram's research (besides his own), and many articles have been written about it. Some of the responses deal with the ethical issues raised both by Milgram's pessimistic conclusions about the deficiencies of human nature and by the method of the experiments, which involved deception of the experimental subjects as well as subjecting them to great distress.

More remarkable than the academic responses to Milgram's experiments is that they became known, not only to well-informed people who might be expected to know about such things, but to the general populace at large. Dramatic narratives were written and produced based on the experiments, including one by English playwright Danny Abse; another on network television, starring the erstwhile captain of the starship *Enterprise*; and a documentary reenactment of the experiments themselves. There have been popular articles, including an early one in *Esquire* magazine entitled "If Hitler Asked You to Electrocute a Stranger, Would You?"[1] An article in *Psychology Today* (online) published in 2002 points out that Milgram's book has been translated into eleven languages and that social scientists, military instructors, and law professors have written or taught about Milgram's famous experiments; and further, experiments testing the degree to which human beings will follow repugnant orders have been carried out by other social scientists.[2]

The Famous Experiments

Milgram presented and summarized his experiments in his book *Obedience to Authority*, describing the experiments themselves in detail, plus his own explanation and commentary. Despite the fact that it is necessarily technical in parts, *Obedience to Authority* is a well-written, remarkably clear, and sometimes-compelling book. Revealingly, the very first paragraph puts the problem of obedience in the context of Hitler's destruction of the European Jews, and Milgram throughout his book mixes straight description with commentary that applies his research findings to what he calls "the dilemma of obedience." "Obedience," Milgram says, "as a determinant of behavior, is of particular relevance to our time."[3] Milgram's experiments were designed to study the phenomenon of obedience to orders from authority in an experimental setting, according to the rubrics of empirical science. Volunteers were asked to take part in an experiment in which the use of pain as a teaching technique was to be studied. The role of the volunteer was to be a "teacher" who would administer electric shocks to the "learner" every time the learner answered a question incorrectly. The learner was first given several matched pairs of words by the teacher, and subsequently, the first word of each pair was repeated to the learner, who was expected to repeat the corresponding word. When the learner gave an incorrect answer, the teacher then administered an electric shock by throwing a switch. After each successive incorrect answer, the intensity of the shock was increased, until at some point, the learner complained. This was the crucial part of Milgram's experiment, for the real purpose of the experiment was to see if the teacher would stop obeying and break off the experiment when he thought he was inflicting real pain on the learner. In reality, the learner was an actor who was uncannily good at simulating the outward signs of someone in pain; no electric shock, in fact, was being administered to the learner. The real subject of the experiment was the teacher, who had to make increasingly difficult choices of whether to break off the experiment, in the face of the orders from the scientists who had ostensibly arranged the experiment on one hand and the complaints, protests, and finally pleas and screams of the learner on the other.

The experiments were designed to simplify the complex phenomenon of obedience by matching the strength of obedience against a countervailing factor, in this case, the moral principle that one should not inflict suffering on a helpless person who is neither harmful nor a threat. Since the electric shocks the teacher must administer were graded in thirty levels of increasing intensity, ranging from 15 to 450 volts, when and if the teacher broke off the experiment, it would be at one of thirty discrete levels. "Behavior prior to this rupture is termed *obedience*. The point of the rupture is termed disobedience." This procedure allowed for quantifiable measurement and control of variables. Great attention was paid to verisimilitude, especially regarding administration of the supposed shock. The shock generator had a series of thirty switches indicating increased levels of electric current, with a red light and buzzer sound every time a "shock" was administered. The teachers were given a sample shock, a real one, applied to their wrists and were assured that even though the shocks "can be extremely painful, they cause no permanent tissue damage."[4] The experimenter under whose authority the pretextual learning experiment was being run wore a laboratory coat, assumed a professional manner, and used a pre-established series of verbal prods to encourage teachers when they resisted shocking the victim. Special attention was paid to the manner in which the learner answered the questions. He gave approximately three wrong answers for every correct one and at the 75-volt level began a preset series of responses, starting with grunts and proceeding to complaints, then demands to be released, then refusal to cooperate, and finally "an agonized scream."[5] At the conclusion of the experiment, each teacher was debriefed and told that the learner had not, in fact, received any shocks.

Milgram ran three series of experiments involving dozens of subject teachers, a process that allowed him to extensively vary the experimental conditions and look for possible causal factors. According to Milgram (and to most anyone who comments on them), "the results of the experiment are both surprising and dismaying," since almost two-thirds of the teachers could be classified as "obedient subjects" who gave the required shocks to the learner to the end. The first point Milgram makes about the untoward results of his experiment is that there is no avoiding them, since the teachers were not sadists and were a fairly representative sample of the population at large. From this, Milgram concludes that the moral sense of the individual is easily overcome by social pressures, and that ethical rules such as the fifth commandment against killing are not an inherent part of human psychic structure. On the other hand, the individual does not lose his moral sense when following orders that contradict his conscience. Rather, Milgram says, his moral concern shifts from his act itself to how well he does his job, wanting to live up to the expectation of the authority figure. Further, he looks to the wider context of the purposes of society for legitimating his actions, blaming the learner for the punishment inflicted on him due to his own stubbornness and stupidity. Milgram points out that evil acts commanded by social authorities today are fragmented in such a way that no one person is ever fully responsible for them, which reduces responsibility for evil acts, a phenomenon which is the "most common characteristic of socially organized evil in modern society."[6]

In *Obedience to Authority*, Milgram spends the last six of the fifteen chapters that make up the book on analysis and interpretation, attempting to provide a theory to explain why the experimental subjects obeyed the commands of the experimenter, knuckling under to authority and performing actions that were callous and severe. Milgram's analysis is of a high level of generality and is complex. He starts by noting the survival value of hierarchy, there being evolutionary (i.e., survival) values associated with hierarchical social organization. From this evolutionary viewpoint, Milgram proceeds to a "cybernetic viewpoint," which will provide a model to alert us to the changes that logically must take place when hitherto autonomous entities are brought into hierarchical structures. Automata living in isolation can be described by using a "homeostatic" model in which the automaton is an open system requiring input from its environment to maintain its internal state. A lack or need felt within the automaton makes it act on its environment to fill that lack or need. An automaton, when brought into a social organization, needs to be regulated if it is not to treat other automata as part of its environment and attack them; hence the need for "an inhibitor that prevents automata from acting against each other."[7] In human beings, this inhibitor is their conscience.

A more powerful form of social organization can be achieved by a hierarchical ordering in which there are subordinate elements controlled by superordinate ones. However, Milgram states, an inhibitor or conscience is not sufficient to allow control of one automata by another, so the human conscience itself must be secondary to the need to cede control to the superordinate element. Because of the variability of individual organisms, each must undergo internal modification, so that all individuals operate at the same socially ordained level. The basic reason for the modification is not rooted in individual needs but in the needs of the hierarchical organization.[8] The process of internal modification that takes place when an individual enters a hierarchical social arrangement that directs his behavior toward obedience, Milgram calls "the agentic shift"; and the attitude of a person acting no longer as an autonomous person but as an agent executing the wishes of his superior, Milgram calls "the agentic state," which is "the master attitude from which the observed behavior flows."[9] The agentic shift is explained in terms of a two-valued logic, and Milgram asks, "[w]here in a human being shall we find the switch that controls the transition from an autonomous to a systematic mode?" His answer is that "[h]ierarchical inhibitors and disinhibitors alter the probability of certain neural pathways and sequences being used," in effect stating that a chemo-neurological cause explains the internal psychological state wherein a person is obedient.

Milgram then applies his theory of the agentic shift to his experiments, arguing that the general consequence of a person entering the agentic state is that he becomes "something entirely different from his former self, with new perspectives not easily traced to personality."[10] The reason is that once a person is in an authority situation, he is subordinate to an authority figure and tunes in to the signals emanating from the authority figure, for the authority situation has redefined the circumstances and details in such a way that effectively dictates reality for the person. As a result, the person experiences a loss of personal responsibility for whatever action he is commanded to perform, for he is now instead concerned with how well he performs in the authority situation. The person's self-image, Milgram says, is no longer involved with the actions he is commanded to perform. The agentic state is not just another word for obedience, for it is that state of mental organization that enhances the likelihood of obedience. Obedience is the behavioral aspect of the state.[11]

To apply his general theory of obedience to the results of his laboratory experiments with real human beings, Milgram introduces the term "strain," which he explicitly uses to replace "choice." Strain explains why some subjects were obedient and some were not. From the cybernetic viewpoint, strain arises because the demands on an autonomous entity are different when a person is by himself than when in a hierarchical situation; that is, the "design requirements of an autonomous unit are quite different from those of a component . . . designed for systems function."[12] In effect, a design compromise has been reached, and the compromise does not always work very well.

Strain is seen in actual sense in the tension experienced by the experimental subjects. The reason they felt tension is that transformation to the agentic state is only partial for some persons, and unlike the more potent authority systems of totalitarian governments, the authority system of the laboratory is less pervasive, allowing "residues of self-hood" to remain intact beyond the experimenter's range of authority. Features that reduce the connection between the subjects' actions and their consequences are called "buffers." Inflicting pain at a distance is easier than when face-to-face, and therefore, "nothing is more dangerous to human survival than malevolent authority combined with the dehumanizing effect of buffers."[13] There are several mechanisms for resolving strain—which were exemplified by the experimental subjects—including avoidance, denial, physical conversion (tics, laughter, etc.), minimal compliance (pressing the switches for a short period), subterfuge, searching for reassurance, blaming the victim, and non-instrumental dissent (i.e., protesting but obeying anyway). All of these mechanisms, however, allow the authority relationship to remain intact. Disobedience is the ultimate means of resolving tension, and while it destroys the experiment as far as the subjects experience it, disobedience is nevertheless a positive act.

Milgram provided answers to likely and actual criticisms of his experimental view of obedience to authority, among them that the subjects obeyed because the scientific setting allowed the release of their latent aggression. Milgram denied the validity of this Freudian alternative explanation, stating that obedience itself, and not aggression, is the key to why men kill in war, for example. He cited experimental evidence that when the teachers were allowed to choose their own shock level, very few went beyond the learner's first protest. In answer to the charge that people who volunteer for psychological tests are a self-selected example, Milgram cited a study that showed that people who volunteer for psychological experiments are less "authoritarian" than those who do not. Finally, in answer to the criticism that the laboratory setting is so special that nothing can be inferred from it about real-life situations, Milgram responded by stressing that however dissimilar the laboratory may be compared to living under the Nazi regime, the psychological process of obedience is invariant. In authority situations, composed of subordinate and superordinate roles, the person responds not because of the content of what he is ordered to do but because of his relationship to the authority figure. For Milgram, obedience to authority lies at the heart of the grossest evils of our time and the consequent moral dilemmas they produce.

Methodological Objections

Milgram portrayed his experiments as explaining the Nazi phenomenon at a time when world attention was focused on the capture, trial, and execution of Eichmann. The tie-in of his experiments with Nazism is made so frequently in *Obedience to Authority* and in popular presentations that it becomes impossible to see his experiments as anything but an anodyne, an attempt to relieve the pain of the memory of the vast evil of the Nazi holocaust of the Jewish people; Milgram mentions the Holocaust in the very first paragraph of his book, as if to say "once we understand or can explain it, then we may prevent it from ever happening again." In a private letter written long before he ran the experiments, Milgram expressed just how much the Holocaust was on his mind: "I should have been born into the German-speaking Jewish community of Prague in 1922 and died in a gas chamber some 20 years later. How I came to be born in the Bronx hospital, I'll never quite understand."[14]

Another reason for the interest in Milgram's experiments is that they blatantly reveal a fact of human nature that we would deny, namely how easily we are influenced by others. It is noteworthy that most people, upon hearing of Milgram's experiments, are repelled by them. Denial of their validity or egregious underestimating of the number of subjects who would obey were typical responses, and Milgram was no doubt correct when he maintained that objections about the ethicality of his experiments would not be half so vehement if most of the subjects had quit them. Our concept of human nature, from the time of the Enlightenment, is that we are autonomous and rational, ruled by our own reason. Lately, we have found out differently, but enough sense of our personal independence and individual reason survives in our culture that we are shocked whenever the discoveries of psychology and sociology contradict it. The irrational inner urges that direct our personal behavior and the fact that social pressures determine most of our opinions are things we do not wish to know. That persons would obey another person simply because he appeared in the guise of an authority is simply something that we would rather not recognize, but Milgram's research has shown it to us, and in our shock and embarrassment lies also its compelling interest.

The reluctance of many observers to accept the finding of Milgram's experiments led to criticisms denying their validity, and while Milgram dealt with several of them in his book, significant questions remain. While we may readily admit that, as Milgram argued, his experiments are applicable beyond the laboratory and that most of his subjects did believe that they were administering a real shock, some lingering questions remain about the selection process by which volunteers were garnered for the experiment. Did they not, in effect, constitute a self-selected pool more ready to accede to authority than the population at large? Surely, people who volunteer for whatever reason will more readily go along with whatever is demanded of them than those who might be forced. That is why particularly dangerous military missions are sometimes put on a volunteer basis and why elite forces that are required to perform such missions—such as the Marines and the Navy SEALs—are volunteer forces. Milgram denied this objection, but the fact of the participants' volunteering being the critical point that sets up the experiments is beyond experimental control or observation. Because Milgram cannot say what effect the act of volunteering has on his experiments, it remains a point for conjecture and speculation.

There are several other significant methodological difficulties that tend to weaken Milgram's conclusions, one of which deals with the question of what constitutes disobedience. Is it simply the act of breaking off the experiment and refusing to participate further in it, as Milgram would have it? Clearly, the subjects gave many other signs of distress and extreme tension; they attempted to subvert the experiment by giving lower shock levels than demanded when the experimenter was out of the room, they gave shocks as brief in duration as possible, they gave verbal clues to the learner, and they argued with the experimenter. Yet, while Milgram carefully described all of these actions, he classified them as "non-instrumental" and did not include them in his measure of disobedience. In real life, however, such "non-instrumental" acts can lose someone his job—or even his *life*, as they did the Russian poet Mandelstaam, who was sent to the gulag because of an obscure reference to Stalin. Surely, if such actions count in real life, they ought to count in some fashion in Milgram's experiments, especially as he claims they are applicable outside the laboratory.

Milgram explicitly discounted many manifestations of resistance to authority that occurred during the experiments, since the teachers continued to administer the shocks to the learner. These were the so-called non-instrumental manifestations, such as verbal protest, not pressing the switch at the next-highest level of intensity, giving the learner clues, or pressing the switches for as short a time as possible. Had each of these non-instrumental manifestations been given a relative value and been included as evidence of resistance to authority along with "disobedience" itself (again, here, the act of breaking off the experiment), a much different picture would emerge, one more subtle and hence more applicable to the complexity of authority relationships in the real world—and one more hopeful about the potentialities of human nature. For instance, each subject could have been assigned a "score": "10" for disobedience, "5" for verbal protest, "2" for not pressing the proper switch level, and so on. Then a curve or sliding scale would result, with relative levels of disobedience, rather than a "sheep versus goat" phenomenon. But such a procedure did not fit Milgram's understanding of how authority works.

Another question arises over the experimenter's statement made to all the teachers that no "tissue damage" would result from the shocks, which amounted to a guarantee that the experiment was not physically dangerous to the learner. According to Milgram, this statement was made only to lend authenticity to the experiment, yet it operates not merely as a detail of Milgram's stage management, like the lights of the shockboard, but also as a factor in the rational decision of at least some of the subjects to remain obedient. We know this because they said so in their statements, which are quoted in the interviews Milgram provides. For instance, the "slow" drill-press operator who showed far less tension than the usual subject cited it as a reason why he continued to press all the switches. He also, not incidentally, himself had once experienced a large electric shock on his job that, as he explained, had caused him pain but not killed him. The point is that the presumed malevolence of an authority and the immorality of obedience to its command are of less degree if carrying out the order causes temporary pain but not permanent damage (a point that was made to defend "water boarding" during terrorist interrogations). Significantly, the statement about "tissue damage" was repeated to encourage recalcitrant teachers, even though it is not listed as one of the four "verbal prods" of increasing rhetorical severity to be used by the experimenters when the teachers refused to go on pressing switches. This, in turn, implies that Milgram himself understood, but chose to ignore, the importance to the teachers of the guarantee that their actions would not permanently injure the learner.

A further methodological issue has to do with Milgram's discounting of the results in the third set of experiments he carried out, described in his Chapter 6. In the first experiment, a stern experimenter and an affable learner were replaced by a friendly experimenter and a tough-looking learner. In the second one, Milgram set up a contract between the learner and the teacher, in which the learner could discontinue the experiment due to his "bad heart." In the third, the experimental setting was changed from the campus of Yale University to a downtown location. All three experiments resulted in significant decreases in the obedience rate: from 65 to 50 percent in the first, from 65 to 40 percent in the second, and from 65 to 47.5 percent in the third. Yet Milgram claimed that these reductions were a matter of statistical indifference, having "little effect" in the first, of "chance variation" for the second, and "not significantly lower" for the third, even though the first showed a 23 percent drop in the rate of obedience as defined by Milgram, the second a 38 percent drop, and the third a 27 percent drop. In previous experiments, variations of the same range or less were considered significant, as in the initial set of experiments described in Chapter 4 of his book, which set the baseline of 65 percent as the standard rate of obedience. These effects, particularly from the contract with the learner, weaken the strength of Milgram's general conclusion about the willingness of ordinary people to obey malevolent authority, especially when combined with the effect of the statement made to the teachers that no tissue damage would result from the experiments.

The net effect of considering the methodological objections to Milgram's experimental view of authority is to weaken it but not to invalidate it. That is, while it is true that people are far more compliant to authority (and social pressure in general) than we might like to admit, the situation is not as severe as Milgram concludes. As we have seen, Milgram, in some measure, must discount the results of some of his experiments, ignore certain phenomena, and define disobedience in restrictive terms in order to sustain the degree of menace he discerns in people's relationship to social authority. His statements constantly placing his experimental results in a political context, particularly in connection with the Holocaust, make it apparent that Milgram had a political as well as a scientific reason for his view of authority. Milgram analyzed a problem, which was how the mad horrors of the twentieth century could have happened—the police state, the gas ovens, the gulag, the atomic bomb, the Cold War (and now the "war on terror")—and reached the conclusion that mankind is far too compliant with authority, which also implies a solution that men ought not to obey authority as readily as they do. The problem, he asserts starkly, is not authoritarianism as a mode of political organization, "but authority itself."[15] But such an analysis either says too much or too little; too *much* because, as Milgram admits, we cannot live without authority; and too *little* because authority is so diffuse and pervasive that that we cannot attack it without further refinement of the issue. Thus, Milgram has, through his experiments, presented us with a dilemma from which we cannot escape. The result of taking what Milgram described as "an experimental view" of obedience to authority is to condemn human nature for its inherent weakness while at the same time encouraging the human race to a task that is beyond its comprehension or strength.

Ethical Objections

Another type of response to Milgram's series of experiments, which appeared even before his book did, were ethical objections made by Milgram's own colleagues in social psychology. It is apparent from Milgram's own description, as well as from films made of some of the experiments, that the teacher-subjects undergo severe emotional stress. Milgram described this effect in an earlier paper, in a passage that did not appear in his book.

> Many subjects showed signs of nervousness in the experimental situation and especially upon administering the more powerful shocks. In a large number of cases the degree of tension reached extremes that were rarely seen in sociopsychological laboratory studies. Subjects were observed to sweat, tremble, stutter, bite their lips, groan and dig fingernails into their flesh. These were characteristic rather

> than exceptional responses to the experiment. One sign of tension was the regular recurrence of nervous laughing fits. . . .The laughter seemed entirely out of place, even bizarre. Full blown uncontrollable seizures were observed in three subjects. On one occasion, we observed a seizure so violently convulsive that it was necessary to call a halt to the experiment.[16]

Such signs of psychological discomfort are obvious indications of severe mental stress. Several psychologists at the time condemned the research for this reason, especially Diana Baumrind, who pointed out that stress and embarrassment were frequent results of current forms of human research in social psychology. Milgram's response to this charge is found in an appendix to *Obedience to Authority* and in an exchange of letters published in Dannie Abse's book, *The Dogs of Pavlov.* He writes here that the degree of stress was not anticipated, but that once he began the experiments, he felt impelled to their completion; and further, that despite their distress at the time of the experiment, the subjects sustained no permanent injury—ironically, the very same justifications used to encourage the subjects to continue the experiment on the *faux* victim-learner. Milgram also argued that many of the subjects said of the experiments that they thought they were good, approved of them, or had learned something about themselves. He also stated that only the subjects may judge the experiment and particularly cited the case of the young man who was a subject and who later refused military service during the Vietnam War.

Whether these arguments justify the degree of stress the subjects underwent is highly questionable, since their effect is to justify the experiment without a cost-benefit analysis of what degree of pain is legitimate (if any!) in human experimental subjects—and whether it depends upon the importance of the experimental results. A cost-benefit analysis, however, would also depend upon whether or how seriously the subjects were damaged psychologically by going through one of the experiments. Here, there is some debate with Milgram's claiming that post-experiment interviews showed no such permanent damage. Yet, one of the subjects described by Milgram, who was "obedient" (meaning, who had pressed all the switches), revealed that after he had described his actions to his wife, she had responded, "You can call yourself Eichmann." Such announcements from a spouse are not easily forgotten. As Baumrind wrote: "From the subject's point of view procedures which involve a loss of dignity, self-esteem and trust in rational authority are probably most harmful in the long run."[17]

Another aspect of the ethical debate is the issue of whether it is legitimate to lie to the subjects in a psychological experiment—in this case, by telling them that the purpose of the experiment was to discover facts about learning rather than obedience and by egregiously misrepresenting the role the subjects were intended to play (i.e., as subjects rather than as agents). Milgram claimed that the deception was legitimate—no different from a play or a movie.[18] But however the final ethical judgment on Milgram's experiments on obedience is made, two points are certain: First, Milgram's most obedient subjects acted no worse than he did. They inflicted (imaginary) pain at the behest of an apparently legitimate authority, while Milgram, in the name of science, permitted the subjects to undergo real psychic pain at the behest of no one at all, despite the protest of his colleagues. Milgram's actions toward his subjects does not have the moral cover that his subjects do, for he acted at the behest of no authority but his own. Second, such an experiment could not be repeated in the United States (and most of the world) today in the same way, under the auspices of a university or a bona fide research institute or with government financial support. The reason is that since the 1960s, when the experiments were run, protections for the human subjects of experiments have been put in place, including disclosure notifications to human subjects and peer-board approval.

Milgram's Scientific Method

Milgram states at the outset of his book that, "Simplicity is the key to effective scientific inquiry,"[19] and indeed this is true. Like Newton, Mendel, and Darwin, Milgram was confronted with a highly variable and epistemologically dense set of phenomena and sought to reduce them to their empirical essentials. Milgram sought to do the same thing, but choosing the elements on which to concentrate is a tricky business, for it is possible to oversimplify in the process of scientific discovery, and this is what Milgram has done. The use of the laboratory setting is, it should be stated, a brilliant stroke, although as he noted, critics have doubted the validity of generalizing from this setting to the world at large. Milgram's problem is that the variety of authority relationships far exceeds the variety of phenomena studied by Newton, Darwin, and such, for the modes of disobedience are infinitely variable, as anyone who has raised an active child can testify. Thus, authority relationships in child-rearing, employment, military, religious, and public-order situations are all very different from each other. While there are resemblances (or else we could not identify them all as authority situations), the differences are tremendous, and the applicability of Milgram's experimental results must be in doubt; unless, that is, Milgram had systematically applied his theories to all these kinds of situations, something he never attempted to do. As Milgram notes, "Psychological matter, by its nature, is difficult to get at and likely to have many more sides to it than appear at first glance."[20]

Milgram's famous experiments yield results typical of much research in the social sciences—theories of a high level of generality that are mechanical and unconvincing; middle-level theories, such as the agentic shift, which are partially successful; and low-level theories or "laws," which are simply inductions from the experimental evidence. Milgram's high-level explanation of obedience is a curious blend of an attitudinal antiauthoritarianism combined with a clumsy amalgam of various mechanistic explanations including evolutionary, cybernetic, neurological, and biochemical. In the end, Milgram contradicts himself with the assertion that "An element of free choice determines whether or not the person defines himself in his way or not," a statement that comes at the end of the Chapter 10, in which he puts forth the evolutionary, chemical, and cybernetic explanations.[21] But an element of free choice relieves the need for "chemical inhibitors" or any comparable explanation in which human behavior is driven by non-intentional causes.

Milgram is on somewhat firmer ground when offering theories of a middle level of generality, including the "agentic shift" that occurs when the individual yields his conscience into the control of an authority figure; which, he states, must happen when an individual enters a hierarchical situation. The notion that a person will perform acts under the command of authority that he or she would refuse to do otherwise is a perfectly valid inductive generalization from Milgram's experiments, albeit one that was known beforehand. This generalization is the basis for the theory of the agentic shift and roughly comparable to Mendel's "genes" in its explanatory function. However, while there is some basis for supposing that such a psychological phenomenon as the agentic shift exists, and while Milgram describes it in some detail, there is no question that he pictures its operations, which are too black and white. That is, before the agentic shift occurs, the person is, in effect, a fully independent and rational agent in control of his faculties, who would refuse to perform any act contrary to his conscience. After the shift occurs, this same person is now ready to perform acts that violate both his conscience and socially accepted moral rules regarding the treatment of other people. There seems no room for the qualified assent of the sort that is often the case when adults obey authority, or for understanding the different kinds and degrees of obedience one gives, for instance, to traffic officers, parents, government officials, bosses, or spouses. The ancient Stoic obeying the laws of Rome, the medieval nun obeying her abbess, and the college-football player obeying his coach exemplify the same empiric law, for Milgram, as the actions of a guard at Auschwitz.

What is lacking in Milgram's theoretical explanations is the capacity to explain satisfactorily the nearly infinite degree of the kinds of human relationships in which obedience is the chief factor, for the authority relationship is not the same in all social hierarchies, nor is it even the same within a *single* hierarchy. Thus, Milgram says that the obedience of the German soldier to his sergeant follows the empirical laws that describe the obedience the Nazi generals gave Hitler.[22] Yet this is not true, for we know that some general officers not merely obeyed like automatons, but aided and advised Hitler, willingly carrying out his orders and proactively executing his policies, while others did not. General Jodl, Hitler's chief of staff, was tried and hanged at the end of the war, whereas Admiral Doenitz, who was Hitler's designated successor, was sentenced to a relatively short prison term. Finally, it was Prussian military officers such as Count von Stauffenberg—the very archetype of authoritarians—who tried to kill Hitler, which shows, if nothing else, the variability of the agentic shift.

Milgram was successful in discovering low-level theories or "laws" because his experiments clearly highlighted the effect of such causal factors as proximity to the victim, closeness of the authority figure, divided authority, institutional surroundings, the nature of commands, and peer rebellion to authority. His experiments also show, implicitly but no less clearly, how powerful authority based on science is. One unanticipated result of Milgram's research is to have demonstrated experimentally that scientific authority takes its place alongside political, legal, military, and religious authority in modern society. It is unlikely that the volunteers would have subjected the victim to presumed electrical shocks at the behest of a minister, policeman, or military officer in that context, coming in off the street, as it were.

That is, while people will perform difficult or even cruel tasks under, for example, military authority, the indoctrination into military life requires an intensive period of "basic training" and the taking of an oath, in a process that takes months. But this is not true of scientific authority in the case of Milgram's experiments, for the teachers gave the apparent shocks after only a minimal indoctrination.

Milgram's Concept of Authority

Milgram never explicitly defines his concept of authority, presuming instead that an experimental view will be morally neutral and present the issue in contemporary and relevant terms.[23] Milgram's actual views about authority and obedience are not difficult to discern, however, for there are attitudinal elements throughout the commentary and explanation in Milgram's book (and in other writings) that very clearly intimate that he perceived obedience to authority as a severe danger to contemporary civilization, that he viewed the exercise of authority by a social agent as often destructive, and that he equated the systematic use of authority with a form of fascism. His many references to the Nazis and the concentration camps, which appear throughout *Obedience to Authority*, including his particular statements about the all-too-human tendency to obey malevolent authority, make his attitude apparent.

A revealing comment appears in his statement about one of the experimental subjects, a professor of Old Testament theology. This subject had stopped giving the victim shocks shortly after the first protest, thus standing as one of the disobedient subjects. Asked afterwards how best to strengthen resistance to inhumane authority, the professor said, "If one had as one's ultimate authority, God, then it trivializes human authority."[24] This answer, worthy of Socrates, Thomas More, or Dietrich Bonhoeffer, is not sufficient for Milgram, who comments, "Again, the answer for this man lies not in the repudiation of authority but in the substitution of good—that is, divine—authority for bad." What Milgram apparently seeks is the destruction of authority itself: "For the problem is not 'authoritarianism,' as a mode of political organization or a set of psychological attitudes but authority itself."[25]

Although Milgram concedes the need for authority,[26] he does not acknowledge how deep that need is: how important role models, parents, teachers, and social authorities are in the formation of our characters as children, or how, even as adults, we still need to look to bosses, peers, spouses, public figures, and religious leaders for direction and support in the daily business of conducting our lives. As a result, he has egregiously misdiagnosed the basic cause of the evils of the modern era, for it can be plausibly argued that the real evil of our time is not that mankind too willingly obeys evil authority, but that we are not willing enough to obey proper authority. As against the examples of wartime atrocities committed at the command of superior officers, we have the large and permanent increase in rates of violent crime since the 1960s and a seemingly permanent culture of rejection of parental and other forms of authority, leading to serious breakdowns in our sense of community and sense of obligation to each other and to the law. The issue of social authority and the degree to which we do and should obey it is far more complex than Milgram acknowledged or implied. Sociologist Robert Nisbet, who was Milgram's contemporary, argued that it is the destruction of intermediate social authorities that stand between the individual and the state that is the main cause of the evils of modern times: "We are prone to see the advance of power in the modern world as a consequence, or concomitant, of the diminution of individual freedom. But a more useful way would be to see it in terms of the retreat of authority in many of the areas of society within which human beings commonly find roots and a sense of the larger whole."[27] The basic human requirement for authority and authority structures that Nisbet points out must somehow be balanced against the dangers of misuse of authority that Milgram points out. This is the true dilemma of authority in our time.

Milgram's attitude toward authority directly influenced his attempt at explaining and evaluating his research results, for he began with a highly charged, if somewhat indefinite, concept of authority and then set up his experiments and conducted his research guided by that attitude. As we have seen, Milgram began with the idea that obedience to authority was a phenomenon not only mixed up with the vast evils of the twentieth century but as the primary cause of those evils. The fact that Milgram had a concept or attitude about the subject he was studying, which then influenced the manner in which he ran his experiments and interpreted their results, is not unusual. Such a procedure is characteristic of the social sciences, where the phenomena are complex and charged with meaning before the scientist even approaches them. The social scientist, no less than the layperson, has ideas about such things as war, politics, economics, family life, and obedience to authority, which, when articulated, constitute a *de facto* theory in scientific terms. Contrary to Bacon, the scientist's mind is not a *tabula rasa* in which explanatory conceptions arise only after he or she has examined the research results. Scientific observation, in Hanson's well-known phrase, is a "theory laden" undertaking, for observation of a particular object or fact in an experiment is shaped by previous knowledge of that fact or object.[28] The priority of the theoretical applies to *all* fields of empirical science, not just the social sciences, although it stands out more in the social sciences because the prior theoretical commitments are likely to be ideological or political in nature. But this highlights a fundamental problem for Milgram and for social science in general; how much assurance can Milgram, and social scientists in general, give that their research results and the theories explaining them are objective?

The Authority of Science

One aspect of authority that Milgram misses is the way that science operates as a kind of authority in contemporary society. It is deeply ironic that Milgram does not comment on this fact anywhere in *Obedience to Authority,* because the authority of science is the effective base of his experiments. That is, the motivating "force" that moved the experimental subjects to "shock" the victim, and the authority against which they were to rebel in order to generate a quantifiable measurement of disobedience, was the authority of empirical science. Indeed, Milgram was meticulous in reinforcing the impression that the subjects were helping carry out a scientific experiment, including such details as white laboratory coats, the electric lights on the shockboard, the text of the newspaper advertisement soliciting test subjects, the laboratory setting and, above all, the actual indoctrination at which the subjects were told that they were helping carry out an experiment. It was against this carefully stage-managed array of cues to scientific authority that the subjects were to rebel, or with which they were to comply.

The use of science as the source of authority was in no way to accidental, since it was the one authority that Milgram could most easily manipulate and the one that was nearest to him. As a bona fide research scientist, Milgram is an authority figure on a par with a minister of the Gospel, a high-ranking military officer, or a police officer, and thus could influence—not to say, *command*—people's behavior in an experimental context. So deeply does the authority of science root in the mores of our society that only a cursory indoctrination was necessary to involve the subjects and get them to perform an act that no other kind of authority could so quickly and so easily motivate—that is, inflicting pain on an innocent person. The reason that science was so effective did not rest, ultimately, on anything Milgram himself did but was the result of a long tradition in the West of accepting science as an authority, which extends back to the Enlightenment of the eighteenth century. Science was, so to speak, the avatar of reason, the one human activity that could produce intellectual certainty and whose practical results could improve the lot of mankind. Science and reason, so conceived, arose from the release of the human intellect from the fetters of religion, metaphysics, superstition, fear, bigotry, and passion—in short, from all the forces of irrationality. Science was seen to be disinterested and benevolent, a model as well as a method for addressing human nature and society. The widespread success of science in the elaboration of theories, in the number and importance of discoveries about the physical universe, and in enabling humanity to control its environment reinforced its status as a source of authority in our culture. It was on this tradition of authority that Milgram depended for effecting the obedience of his experimental subjects.

Two contradictions result from Milgram's use of science as the source of authority in his experiments. The first is that it weakens the value of the very authority, namely science, by which Milgram sought to prove something about obedience to authority. That is, Milgram sought to prove that human nature is too compliant to authority by means of scientific experimentation—at the same time that the authority he depended on in his experiments was scientific authority. If we then accept Milgram's conclusion, as experimentally proved, that we too readily obey authority, we further ought to question the truth of his conclusion, since it is based on our full acceptance of the authority of science. To sum up this strand of analysis, if we accept Milgram's premises, then we both ought and ought *not to* accept his conclusion, which, in the terms of formal logic, means that at least one of his premises in incorrect. Which of these premises is incorrect is difficult to say, but most likely Milgram has overstated the utility of the scientific method in explaining social phenomena, such as obedience to authority. In less formal terms, the real effect of Milgram's experiments is to weaken science as an authority system, not only because they necessarily weaken *all* authority systems, but also because of the serious ethical issues the experiments raised, namely that he misrepresented the purpose of the experiments to his subjects and put many of them through extreme levels of stress.

The second contradictory result is that Milgram's experiments revealed something about human behavior that is contrary to the rationalistic tradition in which science plays an essential part—that is, that human beings tend to obey authority even when it violates their personal notions of morality; and further, that a person's values and identity are more the result of their social environment than their personal choice. In short, science as a rational method of discovery has determined that human nature is irrational. This discovery seems to have depressed and angered Milgram; however, it is important to note that other research conducted by social scientists suggests that other aspects of human nature contradict Enlightenment expectations: for example, that humans are aggressive and territorial, that they tend to band together in tribes and families, and that their intelligence and gender roles are influenced, if not determined, by their biologies. Science has authority because we expect that it tells us the truth. *Now* science—through the discoveries of psychology, sociology, anthropology, and biology—is reaffirming things about human nature that Enlightenment rationalism denied but that older traditions, such as Christian doctrine and Greek philosophy, affirmed.

Milgram argues that while some critics of the social sciences claim they do not discover anything already known, that is not true for his series of experiments.[29] Yet the facts that individuals are easily influenced by others and often will obey malevolent authority was well known prior to Milgram, they were known even to the ancients. Aristotle defined human beings as the "social animal," and the fact of our social nature was well known to those close observers of life in the Greek city-states, Socrates, Plato, Thucydides, Euripides, and the Sophists. Milgram's "discovery" that human beings will perform deeds under the command of authority that they otherwise would not perform—which includes acts of heroism and bravery, not just cruelty—is but one specific aspect of a general fact now amply demonstrated by social psychology and sociobiology (see Wilson's notion of "sociality" in the prior chapter), namely that individual personality contains large elements that are socially and culturally induced. To repeat, the social nature of human psychology contradicts our cherished Enlightenment illusion that we as human beings are both rational and independent beings capable of being in full command of ourselves, and it is ironic that modern science provides the contradictory evidence.

Notes

1. Paul Meyer, *Esquire*, Feb. 1970; 72 *et seq.*
2. "The Man Who Shocked the World" by Thomas Blass; *Psychology Today* on-line, Mar/April 2002; Document ID 1989. *Psychology Today* has published numerous articles on Milgram's experiments starting with a piece by Milgram in May, 1967; see also articles in June 1974, March 2002, May 2002, Sept. 2009, Aug. 2010.
3. Stanley Milgram, *Obedience to Authority: An Experimental View*; New York, Harper Colophon, 1975; 1.
4. Ibid., 19.
5. Ibid., 23.
6. Ibid., 11.
7. Ibid., 127.
8. Ibid., 131.
9. Ibid., 133.
10. Ibid., 143.
11. Ibid., 148.
12. Ibid., 153.
13. Ibid., 157.
14. Ibid.
15. Milgram, 179.
16. "Behavioral Study of Obedience," *Journal of Abnormal Psychology*, 67, (1963), 371–78.
17. Diana Baumrind, "Some Thoughts on Ethics of Research; After Reading Milgram's 'Behavioral Study of Obedience.'" American Psychologist, 19 (1964), 42123. Quoted in Abse, *The Dogs of Pavlov*.
18. Dannie Abse, *The Dogs of Pavlov*. London, Valentine-Mitchell, 1973; 37–44, 126–127.
19. Milgram, 13.
20. Ibid., 13.
21. Ibid., 134.
22. Ibid., 130.
23. Ibid., xi.
24. Ibid., 49.
25. Ibid., 179.
26. Ibid., 212.
27. Nisbet, Robert A. *The Quest for Community*. New York, Oxford U. Press, 1978; 229.
28. Hanson, Norwood. *Patterns of Discovery*. Cambridge, Cambridge U. Press, 1965; 19. The same point is also made by Popper, Kuhn, Einstein, and Duhem.
29. Milgram, 27.

Part IV

Decline and Consequences

Chapter 11

The Cultural Decline of Physical Theory

Confidence in the scientific model of reason, which has been fundamental in Western culture, rests on the positive assumption that modern science provides the best method of uncovering truth. Recently, however, there has been a discernible cultural shift away from the ideal of scientific reason, reflected both in popular culture and in academic discourse. And while the denigration of scientific reason is something to applaud to the degree that it deflates the hubris of the Enlightenment ideal of scientific reason, the devaluation of science in contemporary culture is having deleterious, indeed *destructive*, effects in the processes of constitutional democratic governance.

Science as Portrayed in Science Fiction

Science fiction has progressed from a genre appealing to a limited audience of teenaged boys and men interested in technology and scientific gadgetry ("nerds") to now include intensely produced and marketed films that have attained a worldwide popularity (such as the *Star Wars* and *Star Trek* series, *Avatar*, and many others). The outlook that these films present is grandiose and essentially positive, making outer space and the future a pleasant and exciting prospect onto which the audience is invited to project its fantasies. Such fantasies are not scientific in any exact sense, however, and are certainly not new, but rather are age-old themes found universally in literature and ancient myths: romantic love, strong-jawed heroes, evil villains whose power can destroy entire planets, temptation and mortal danger, family sagas, adventure and derring-do, unearthly life forms, and monsters. The themes found in *Star Wars* are virtually the same found in Homer's *Odyssey*. However, these films and other forms of what are today loosely called "science fiction" are far from any exact reference to modern science, because science fiction in these forms has become escapist literature or social commentary.[1] But the science fiction referred to in this section is literary and reflects the actual manner in which science is done, what science's discoveries portend for the future, and the place of science in contemporary society. In this manner, it is, in effect, the older tradition of science fiction that is referred to here, which has a richer intellectual content and more permanent cultural influence.

* * * * * *

Fans of classic science fiction from the 1950s must have been pleased when the prestigious Library of America gave credibility to the science-fiction genre by publishing a volume of one of its foremost practitioners. However, that pleasure may have been diminished by the choice of a representative author; rather than Isaac Asimov, Arthur C. Clark, Robert Heinlein, or Ray Bradbury, true giants of the form, the editors of the Library of America chose Philip K. Dick instead, whose *oeuvre* was largely characterized by dark visions of the technological future. But the darkness of Dick's vision was not based on apocalyptic descriptions so much as on a vague-but-persistent paranoia about social organization in general. This Kafkaesque theme, of the individual placed in a setting both familiar and alien, apparently appealed to the editors of the Library of America because it is reflective of the cultural suspicion of contemporary forces such as not only government and military but also large corporations whose effective area of influence and control has been tremendously enhanced by postindustrial developments in technology, including the internet, ubiquitous digital cameras, and genetic alteration.[2]

The movie *Blade Runner* was based on Dick's novel *Do Androids Dream of Electric Sheep?* Its vision of a dystopian future is based on the premise that androids—robots built and designed to act like human beings and indistinguishable from humans—have turned against their human creators to become killing machines. The setting of the movie presents the future Los Angeles as a hodgepodge of Asian food stands, large electric signs that dominate the field of vision, and background noise made up of engine sounds, sirens, and snatches of incomprehensible conversations—a constant stream of unconnected, yet vivid, impressions; but its most disconcerting feature is the omnipresence of the police, who float above the city scene in flying machines, and the undercover cops, who make unannounced appearances. This film version of Dick's vision of the future has become the standard of much futuristic film, a dystopia more frightening than the post-apocalyptic version in the *Mad Max* film series.

The *Matrix* series of science fiction films, featuring the hero known as "Neo," presents a vision of the future that spells out the paranoia endemic in contemporary culture, explicitly based on scientific technology. For technology and, especially, computers have made it possible that what we experience as everyday life, our social world, is a sham and unreal, for it is a creation by unknown and dark forces. The unarticulated unease in Dick's dark vision is reduced in the *Matrix* series to an apparatus by which Neo's brain is literally wired—the top of his skull removed and electrodes attached to the exposed brain mass—and the conflict made explicit with the evildoers wearing black suits, as in an old-style western movie. The action scenes are extreme, taking place in slow motion in which the characters are seen weaving their way through space, avoiding the bullets flying at them and performing acrobatic moves that are impossible except when enabled by the application of electronic, pixilated movie art. In the *Matrix* series, the idea is that scientific technology is the chosen weapon of deniers of freedom, who constitute an overriding threat to the individual and who must be mastered by defenders of the right against, by implication, evil government forces.

That the Romantic tradition is alive and well in contemporary popular culture is seen most prominently in the character of Mister Spock, who appears in the *Star Trek* television series, which is both a cult television program and the basis for a series of popular movies. Spock serves as the emotionally reserved second-in-command aboard the starship *Enterprise* and who stands in dramatic opposition to the ship's emotional doctor; Spock's character provides a vivid manifestation of the Romantic view of the scientist—uninvolved, rational, and detached, because he is the bearer of secret knowledge. Spock is the starship's "science officer," a noteworthy title since it designates not a function—such as the navigator, who must surely know the science of astronomy, or the ship's communications officer, who must know electronics—but a more general role, that of the representative, and very symbol, of scientific knowledge. The tension in the episodes frequently involves Spock wrestling with his own emotionality in an argument with the emotional physician or in conflict with the energetic captain who, unlike Spock, knows when it is time to stop thinking and time to act.

The inability of Spock to act in critical situations is an indication of a new development in the West's understanding of modern science. Hitherto, science, as portrayed in the Romantic tradition, was an unstoppable force, cunning in its way of giving mankind the ability to penetrate the inmost secrets of nature and of life while overwhelming the resistance of religious belief and received tradition. The most recent attitude, however, seems to be that while science gives us the knowledge by which to control nature, that power is limited, because science cannot tell us to what ends to direct our newly acquired control. That is, modern science gives mankind the physical ability to manipulate nature but does not provide the ethical concepts by which to guide our actions. Hence, Spock's inability to decide or act despite his superior scientific knowledge turns him into a *Hamlet* character who, just because of his immense knowledge and objectivity, has lost his power to choose. In this, Spock resembles the atomic scientists at Alamogordo, whose knowledge and skill developed the atomic bomb but who had no say in whether or how it was to be used. The inability of scientists to extend their knowledge from their own fields to the arena of practical ethics was never more apparent nor more tragic than in the development of atomic weapons.[3]

The invalidism of modern science implied in the character of Mister Spock is more explicit and more malignant in the science fictions writings of C. S. Lewis, the notable fantasist, literary scholar, and Christian apologist. Among his works is a space trilogy consisting of three novels, whose action takes place, in order, on Mars, Venus, and Earth. In these novels, Lewis sharply distinguishes between the power that science gives mankind over physical nature on one hand and science's inability to provide valid ethical guides for the use of that power on the other. Lewis's novel *That Hideous Strength* (the last of the three, published in 1945) describes the actions of a scientific cabal to take over England, which uses the full panoply of totalitarian techniques, including propaganda, corruption of government officials, police brutality, and brain washing.[4] The hero is a naive sociologist who is torn between the forces of the cabal and its opponents, a collection of English eccentrics; this odd group, which enjoys the help of angelic beings, eventually includes the professor's wife, a resuscitated Merlin, and other odd characters. The novel uses Christian and classical pagan symbols as an attack on the illegitimate use of science for the purpose of social control. In its depiction of scientists and the scientific project as evil, the novel probably has no equal.

Significantly, the focus of Lewis's attack is not the hard sciences but the social sciences, which he thought illicitly imported scientific techniques into areas traditionally the province of the liberal arts. In *That Hideous Strength*, one of the victims of the scientistic cabal is Hingest, a physical chemist who thinks that what the cabal's "Institute" is up to is politics, and not science at all. "I came here [to the Institute] because I thought it had something to do with science. Now that I find it's something more like a political conspiracy, I shall go home." To this, the sociologist replies, "You mean, I suppose, that the element of social planning doesn't appeal to you? I can quite understand that it doesn't fit in with your work as it does with sciences like Sociology." Hingest interrupts: "There *are* no sciences like Sociology."[5]

Lewis's distinction between the hard sciences and the social sciences is important, because it indicates that the real basis of his mistrust was the misuse of science by importing its methods into the study of human beings. He has Hingest, his spokesman, say, "That's what happens when you study men; you find mare's nests. I happen to believe that you can't study men; you can only get to know them which is quite a different thing." The question for Lewis is the essential ethical one: how are we to live, and what rules are there to follow? Lewis's answer is revealed religion in the form of Christianity and the collected wisdom of the human race; he sees science as an intruder into the field of morals and social ethics, spreading not insights but only confusion and mischief. Lewis's objection is overstated, for surely science, including social science, has told us some things about human nature that we did not understand before—at least not with the definitive clarity that scientific method can sometimes bring. Nonetheless, Lewis's point reflects a set of moral dilemmas posed by science in the twenty-first century.

Recent Academic Criticisms of Modern Science

We are currently in a cultural quandary in which technology continues to expand its domination over every aspect of our lives, from birth-control pills that separate sex from reproduction, to instantaneous electronic communication devices that constantly distract our attention, to hydrogen bombs preset on their rocket-powered, gyroscopically guided path to Armageddon. At the same time, we lack an ethic of scientific practice, a principle that will contradict the innate sense that if it is *possible* to accomplish something technically, we simply *ought to do it.* The atom-bomb scientists of the Manhattan Project, including Oppenheimer and Teller, referred to the solutions of certain scientific problems that cropped up in the development of nuclear weapons as "technically sweet," an instrumentalism that obscured the requirement of moral evaluation.

An implicit instrumentalism seems to come along as an inevitable partner of technical development, making us somehow unable to slow down the pace of our lives—which have been speeded up by fax machines, cellular phones, and the World Wide Web—or to slow down the rate of heart attacks and abortions. A recent example is the controversy over infant stem-cell research that took place during the administration of George W. Bush. The creation of embryos by means of *in vitro* fertilization, which are then disassembled—that is, *killed*—and used for experimental research would seem, on its face, to present a moral issue. But the putative benefits simply seemed to overwhelm the moral considerations, such that when Bush attempted a compromise on the issue, he was roundly condemned as being "anti-science." In the infant stem-cell controversy, the instrumentalism of medical technology is most apparent; also notable is the fact that the results of such research were wildly overestimated. (As regards the moral issue, infant stem cells are distinct from stem cells obtained from birth cords or from adults, because in the latter, obviously, fetal life is not destroyed; see Chapter 14.)

The same motive of instrumentalism has inspired the environmental or "green" movement, which, despite its extremes, is largely a response to the environmental evils connected to the centuries-long process of western urbanization and industrialization. The disappearance of entire species, pollutants in the air, and toxic chemicals in the water supply are examples of the unwanted effects that the application of science has had in contemporary Western culture. The environmental movement reflects a deep unease, a consciousness that scientific progress comes at a cost and that considerations external to science and its technological applications have to be consulted in order to set social and individual priorities.

The crises of social and individual conscience provoked by new means of scientific control have led to a profound conviction that scientific progress is not an unalloyed good. As a result, the suspicion of science in our contemporary culture has deepened to the point where it is no longer possible to effectively combat it by optimistic visions of the future and declamations about the special place of science in the advancement of human history. No longer does General Electric claim that "Progress is our most important product," and DuPont no longer brags, "Better living through chemistry."

There has been a further development in the general cultural objection to utilitarian science, coming from another, unexpected source; for scientists and defenders of the scientific project are just now becoming aware of attacks from the cultural Left and appear somewhat perplexed. Whereas the hard sciences have formerly been untouched on campuses where political correctness has invaded the humanities departments and the social sciences, the hard sciences have now become targets for political correctness. Scientists generally ignore these attacks, since they do not have an immediate effect on their work. And practitioners of the hard sciences in fact may not have the wherewithal to counter these attacks on the intellectual level, because they are well known for often having little sense of the history of their fields and for having a disdain for historians and philosophers who comment on science.[6] The unwillingness to deal with cultural criticism places science at a significant disadvantage, since the most influential attack on science in our time comes from an entirely new front—not, as formerly, from Romantic poets like Blake and Wordsworth or from historians nostalgic for the Middle Ages or ancient Greece, but from the cultural Left. That is, feminists, multiculturalists, and deconstructionists—what critic Harold Bloom calls, "the schools of resentment."

In 1994, a book written by a mathematician and a biologist sounded an alarm; *Higher Superstition: The Academic Left and Its Quarrels with Science* began a counterattack, although it is not a defense of science but rather an attack on its critics.[7] It is a well-researched book that gives evidence that its authors wanted to cover all the relevant points of left-wing criticism while giving back as good as science has received from these critics. The tone of the book is one of anger and even contempt—for feminists, for example, who criticize the history of physics but cannot do the calculus; multiculturalists who attack science while depending on a daily does of insulin; and deconstructionists who treat scientific-research results as if they are works of fiction. The weakness of the book is that it neither presents a positive case for modern science nor concedes that a case needs to be made, even in the face of such persistent criticism. Instead, *Higher Superstition* presents a face of angry triumphalism—angry, that is, at the smallness of science's critics and the manifest stupidity of their attacks, and yet angry also because it has even become necessary to answer back.

Of the various schools of criticism from the cultural Left, feminist critics of science allege that, as a historical fact, science has relegated females to a subordinate role. They point to the virtual exclusion of women in portrayals of science history up to the last generation or so, and they have attempted to correct the record by writing biographies of female scientists who, it is asserted, did not get the recognition they deserved because of their gender. Included among such female scientists are geneticist Susan McClintock, physicist Liese Meitner, and X-ray diffraction specialist Rosalyn Franklin. Multicultural critiques of the scientific project of the West allege that modern science is just another Western tradition that should not be used to rationalize Western superiority over non-Western cultures, for science has been utilized as a means of cultural objectification of so-called primitive societies and physical control of native populations.

While specific types of criticism have come from feminists and multiculturalists, a more fundamental underlying critique of modern science emanates from deconstructionist critics such as Lyotard and Foucault; these criticisms go beyond citing the prejudices of scientists or the unintended consequences of technology—they deny that science has the ability to know objective truth. The authors of *Higher Superstition* comment that "this high-handed attitude toward science and toward rationality itself has its instrumental uses within the narrow world of left-wing campus politics. . . . If one believes that science is nothing but metaphor, then one need have no qualms about substituting it for other metaphors."[8]

A deconstructionist critique of science typically applies the same set of standards and methodologies that are applied to literary texts to the research results of empirical science. This approach implies that science is seen as an activity that has, as its function, the production of texts for the purposes of establishing reputations and gaining access to power, prestige, and funding for research. What gets left out of the deconstructionist account is that aspect usually supposed to be the essence of scientific activity—namely, the unprejudiced search for the objective truth about the physical universe by means of experimentation and close observation. The deconstructionist approach, however, treats science no worse than, or in the same fashion as, it treats literature, where the interpretation of poems and novels is totally in the hands of the deconstructionist critics rather than those of the author or the readers. But in response, scientific texts in the form of research papers or monographs are notoriously difficult to understand for those outside their immediate fields, including other scientists, so that deconstructionists would seem obstructed from science criticism by their inability to understand the meaning of scientific texts. The fact that such criticism goes on is based on the denial of the worth or validity of scientific texts in themselves, a point that the authors of *Higher Superstition* take up. They assert, not unreasonably, that to be credible, left-wing critics of science should have a solid understanding of the scientific fields they criticize.[9]

What can happen when deconstructionist critics do not understand a scientific field is illustrated by a fraud perpetrated by physicist Alan Sokol in the journal *Social Texts*. Sokol wrote a phony article entitled "Transgressing the Boundaries: Toward a Transformative Hermeneutics of Quantum Gravity," whose very title reflected the point of view of this prominent deconstructionist journal.[10] Sokol's article combined, in effect, the lingo of the social sciences at its loosest with actual scientific research results; and its acceptance, he asserted, demonstrated the outright scientific ignorance of the journal's editors and readers. In the journal's defense, such a mixture of the two terminologies is not very hard to bring off, as interpretations of research results by quantum physicists themselves tend toward high metaphysical flights, with speculations about "alternative universes," "bilocation," "entanglement," "complementarity," "indeterminism," "uncertainty," "weirdness," and, not least, speculation that the very act of perceiving subatomic phenomena creates the phenomena being studied. (See Chapter 8.)

In their defense of modern science, the authors of *Higher Superstition* rely on an exclusionist understanding of science, in the Enlightenment sense, as something apart from and uninfected by usual human concerns.[11] Yet science can be turned upon itself so that *it* becomes an object of scientific study, for science can be seen as an anthropological artifact, a point that advocates of the deconstructionist approach readily understand. The separate fields of science, such as astrophysics or cell biology, can, like the various non-Western cultures examined by anthropologists, be seen as enclosed within themselves, their modes of discourse and means of dealing with the world incommensurable to one another. "Incommensurable," of course, is Thomas Kuhn's term, used by him to explain why representatives of competing paradigms within a scientific field cannot understand one another—and Kuhn thus stands as the chief deconstructionist critic of science.

What is lost in these academic criticisms is the clear understanding of, and belief in, scientific objectivity, the loss of which damages the very motive for doing scientific research. The parallels with multiculturalism are precise and ominous. The existence of separate cultures that are worthy of an equal degree of respect is presumed to mean that they are not to be judged by a common standard (particularly one imposed by Western culture). So, also, since each distinct scientific field is dominated by its own paradigm, and paradigms are incommensurable, there can be no common standard of scientific truth or objectivity that can be applied to them. This relativistic implication of Kuhn's approach to science was notably recognized and opposed by his critics. (See Chapter 7.)

Effect on Democratic Governance

There are sound reasons for thinking that the scientific project is overvalued in modern Western culture, especially given the influence of scientific thought and the effects of scientific technology on culture, combined with the presumption that science is the teller of categorical truth; but there is a downside to the denigration of science in our culture, as well. The Declaration of Independence's famous first words, "We hold these truths to be self-evident," indicate a far greater confidence in the power of human reason than is now often found in Western culture. Jefferson, who wrote the document, was a devotee of the new mechanical science that had arisen in the previous century; and, according to one account, he considered Newton the greatest man in human history. Jefferson's rationalistic approach to the Christian religion is well known; he is considered to be a "deist" rather than a believer in traditional Christian doctrine; he composed a version of the Bible in which Christ's miracles were excised. The reliance on scientific reason as the model of rational thought, typical among intellectuals of the eighteenth century, is no longer apparent in the twenty-first, in part because the main scientific paradigm on which science was based in those days, Newtonian mechanics, has been overthrown; it has been replaced by a bifurcated account of cosmic cause and effect, relying on relativity and quantum mechanics. In short, reliance on scientific reason has been undercut by the progress of scientific knowledge itself.

The Constitution, as well, is a "scientific" document, in that it attempts to impose a rational framework on the political processes of the former colonies of North America by a balance of powers. But a balance of powers is not only a political theory based on the history of republican states; it is also a figure taken from the physics of the eighteenth century. No less than Newton's vision of the planets traveling in their orbits, avoiding collision and establishing celestial harmony, do the arrangements imposed by the Constitution act in their countervailing way to bring about, peace, security, and justice to the American polity.

The rationalism of the federal Constitution is based on less a theory of abstract rights than a hope that the partisan pushes inevitable in a democracy can be set against each other in such a way that no interest, faction, or class becomes all powerful. Since human beings are not planets or cannon balls, their behavior does not follow directly from external forces but rather is guided from within by their own perception of self-interest and the common good. Thus, the Constitutional balance of powers will continue to do its work only to the degree that a workable majority of American citizens agree. The hope is that people will be rational in the precise sense—that is, that their reason will follow, in effect, objective and scientific standards. With the denigration of scientific reasoning necessarily comes a danger to the constitutional order, to the degree that that order is modeled on scientific reason. For when the intellect is no longer respected, the power of naked will and acts of assertion become the chosen means of political activity, while political debate acquires a more destructive tone. Then coercion and crass manipulation become the means of securing political ends, which is destructive to the soul of democracy.

Notes

1. On the distinctions among various types of science fiction in its contemporary forms, and what forms of literature can be termed "science fiction," see Margaret Atwood's dissent from Ursula K. Le Guin, in *Other Worlds: SF and the Human Imagination*; New York; Doubleday; 2011; 1–8.
2. Philip K. Dick, *Five Novels of the 1960s & 70s* (2 vols.); Library of America, 2005. See article by Rob Latham, "The Exegete" in the *Los Angeles Review of Books*, February 24, 2012, a review-essay on Dick's later works in which his paranoia is described in full, psychotic bloom. http://lareviewofbooks.org/18187221884/the-exegete (accessed 2-24-2012).
3. See John Caiazza, *War of the Jesus and Darwin Fishes*; New Brunswick, Transaction, 2007; "The Agony of J. Robert Oppenheimer, 109–118.
4. Clives Staples Lewis, *That Hideous Strength*; New York, Macmillan, 1965.
5. ibid.; 70–71.
6. See Weinberg, *Dreams*; the chapter "Against Philosophy;" 173–198.

7. Paul R. Gross and Levitt, Norman, *Higher Superstition: The Academic Left and its Quarrels With Science*; Baltimore, Johns Hopkins U. Press, 1998.
8. ibid.; 225.
9. ibid.; 235.
10. Alan Sokol, "Transgressing the Boundaries," *Social Texts*, #46/47, 217–252 (spring/summer 1996).
11. Gross and Levitt, *Superstition*; 19–20.

Chapter 12

Three Religious Fragments

The current diversity of religious opinions in America following the decline of what could reasonably be termed "*the* American religion" precludes summary description or simple analysis. Yet, some analysis needs to take place in order to understand the state of religious belief in postmodern culture when an aggressive secularism is on the prowl; but again, the field of inquiry is too varied, and rather than proceeding from point A to point B in a straight, purposeful direction, it is required, apparently, that the analysis wander not only from side to side but from one narrative account to another. Further, religion, as a subject of academic study, includes a tremendous variety of different approaches, including, among others, anthropological and neurobiological, the extent of which can hardly be summarized and comprehended.[1] A more useful approach is to analyze three contemporary forms of religious belief (or unbelief), which are having significant impact and which will, without stretching the matter too much, define the limits of current religious thought "after the fall." Thus, I will deal with contemporary Gnosticism, which attempts generally to find the divine deep within ourselves or as a hidden aspect of human history; evolutionary atheism, which depends on natural selection to explain all aspects of human life, including religious belief; and Pope Benedict XVI's controversial speech to the faculty at Regensburg, which resulted in riots across the Muslim world but whose content elaborates a contemporary case for monotheism.

American Gnosticism

Gnosticism has been analyzed in a broad historical context to explain twentieth-century ideological politics by one prominent philosopher, and by historians of religion and literary scholars who can discern in Gnostic doctrines connection to the most varied items of human experience.[2] While its basic assumptions and doctrines can be somewhat defined, it not easy to do so, because Gnosticism has never been a unified social entity (i.e., a church or a cult or an organization, except possibly the Rosicrucians or the Masons). Its basic thrust is that the human mind can discover a secret that is the key to explaining and understanding all the mysteries of life and the course of human history, a cosmic clue that, once recognized, brings order to chaos and security to one's life.

In terms of religious belief, the Gnostic approach guarantees that God is within ourselves, and thus within our grasp without assistance. "Yet to know oneself, at the deepest level, is simultaneously to know God; this is the secret of *gnosis*."[3] This is an appealing theme, with much resonance in virtually all religious traditions—that to know oneself is a means of contact with the divine. However, the question that remains unresolved is whether self-knowledge reveals, as it were, the essence of divinity with nothing left over, or whether self-knowledge is first required for us to recognize the reality of the divine as a reality external to the believer and not under her control. Gnosticism starts with an insight common to all religions but forecloses the immense possibility that the divine, God as a person or a force or a presence or a design, is something different from and greater than ourselves. Nonetheless, an analysis of Gnosticism is useful, because it is as close as one can get to a coherent description of the current drive for spiritual knowledge and experience without the apparent impedimenta of formal, organized religion.

In the writing of influential literary critic Harold Bloom, who has thought seriously about it for a long time and for whom it is a satisfactory form of religious belief, the appeal of Gnosticism to the literary imagination is apparent. Bloom cites reference after reference, including studies of obscure topics in religious history, pursuing his theme from the Zoroastrians to the ancient Hebrews to the early fathers of the Christian Church; and to the school of Islamic mysticism called "Sufism" to American episodes including, most notably, Emerson in his essay "Self-Reliance"—but also the Mormons, New Age religion and even the futurism of Newt Gingrich. How can Gnosticism, or *any* literary theme, retain any coherent identity when it includes all these manifestations and more? Bloom admits the inherent vagueness of the Gnostic theme: "Yet Gnosticism is an ambiguous term . . . [t]here were, as far as we can ascertain, few, perhaps no Gnostic churches or temples in the ancient world. And yet Gnosticism was more than a tendency, more even than a party or a movement; I think it best to call it a spirituality, one that was and is a deliberate, strong revision of Judaism and Christianity, and Islam later."[4] In the end, however, the real central base of the Gnostic theme is the *self*, and to express this point as fully as he can, Bloom quotes Emerson: "That is always best which gives me to myself. The sublime is excited in me by the great stoical doctrine, Obey thyself. That which shows God in me fortifies me. That which shows God out of me makes me a wart and a wen."[5]

Lately, several authors have taken a Gnostic approach to explaining Christianity. The most prominent figure here is Elaine Pagels, who summarized the new developments based on the newly discovered texts of the Dead Sea Scrolls and Nag Hammadi and on recent scholarship about early sects such as the Essenes, which were contemporaneous with the rise of the Christian Church and Gnostic communities that came after it. Recently, accounts of the origins of Christianity based on non-canonical gospel accounts of Jesus's life have appeared and have an up-to-date appeal, even as they advocate the ancient philosophy of Gnosticism. In the case of the discovery of alternate Gospels, the characteristic Gnostic secrets (now no longer a secret, since it is the subject of best-selling novels, popular books, and influential scholarly monographs) are that the figure of the One God descending, from the Hebrew Bible, is merely the invention of an ancient tribal culture; that Jesus was simply a man; and that, by implication, the doctrines of religious revelation are fully knowable by the human mind.

Interestingly, unlike Bloom and Pagels, who are among Gnosticism's leading scholarly avatars, the individual human mind, relying on this secret fund of knowledge, need not be well educated or particularly intelligent, since the Gnostic secret, in the case of Christian origins, appeals to the basic suspicion of the uneducated that all the things in the world that they do not understand are the result of the actions of the people on top—the "big boys" who run things, whether it be the economy, politics, or religion. These rich and powerful people, it is said, desire above all else to stay in power and so work relentlessly to keep the levers of power to themselves and hide the truth from the rest of us for their own advantage. Here, the big boys have conspired to elevate the simple Galilean preacher into a heroic figure, a god-man, imputing to him all sorts of impossible miracles and wise sayings, an imposture which is to the advantage of those who, following the lead of the ferocious front man St. Paul, playing Lenin to Christ's Marx, erected an authoritarian church with arbitrary rules enforced by threats of eternal punishment to rule the rest of us and enable the big boys on top, now mostly residing in the Vatican and other international venues, to get rich and stay in power. Added to this conspiratorial account is the romance of Jesus with Mary Magdalene, who is changed from the representative figure of holy repentance (in the canonical gospels) into a coy mistress, a narrative that appeals to the romantic imagination; from Pagels's *The Gnostic Gospels* to Dan Brown's *The DaVinci Code* is but a quick step. The general appeal of the recent scholarship of non-canonical gospel texts rests in this crude populism of conspiracy theory and romantic longing.

But despite the vulgarity of *The DaVinci Code*, there remains a nugget of interest, for the appeal of the new Gnosticism rests on a particular view of human reason in relation to divine things, which denies the authority of scripture as revelation and replaces it with human narrative. There is present in Gnosticism a denigration of the notion of revelation, which is the chief characteristic of the Bible and canonical tradition; for revelation provides the guarantee of the Bible's epistemic authenticity. The notion of religious revelation is that God or other divine agencies have handed down truths to the human race, which by and large, it otherwise could not attain. That the Lord God is One and ruler and creator of the universe, that he exists as a Person independently of human understanding, that there are certain ethical rules prescribed for the whole human race in a minimum set of universal ethical standards embodied in the Ten Commandments, and, more particularly, that God formed the Hebrew people for himself, or that Jesus is the true son of the living God, or that Muhammad is God's last prophet are all truths revealed—or *claimed* to be, as these more particular revelations are points of great conflict and contention among the Abrahamic faiths.

But the new Gnosticism denies all this, on one level treating the ancient biblical texts just as texts and nothing more, applying to them no lingering sense of reverence and thus treating them in the same manner as current postmodern analysis treats Joyce's *Ulysses* or Pound's *Cantos*. This new mode of analysis generally treats the texts not as the creation of their authors but rather places the meaning of the text into the reader's hands—or minds. This allows the understanding of texts to proceed as if they are mere products of a specific time and place and of no possible universal import, except as examples of the scholar's particular theses about patriarchy, authority, the suppression of minorities, the danger of politicized religiosity, or the rigidity of sex roles. Applied to revealed texts, necessarily this approach has destructive effects, but it begins with the reduction of revealed texts to simply *texts*, no different from any other. This reduction, in turn, is given credence by the discovery of the Dead Sea Scrolls and other ancient texts that provide an alternate vision of Christian origins based on non-canonical Gospels (e.g., attributed to the apostle Thomas).

Gnostic professors of religious history such as Elaine Pagels and Karen Armstrong now present the face of Christ in a new setting, framing him in a way less patriarchal, less forbidding, and less implicitly authoritarian. But did Jesus really marry Mary Magdalene? If so, what credit to him, for if Jesus had stayed at home with a wife and children, as proposed in the egregious *The DaVinci Code*, who, then, would have ever heard of him, and what would he have accomplished—and who would then be concerned if his later descendents had been the Merovingian kings of France? Reading the Gospel of Thomas, what new revelations are there? Jesus upbraids oafish Peter about Mary Magdalene's place in the new arrangements and makes other striking statements that are either taken from the four canonical Gospels or common to Gnostic thought. But he rarely *does* anything, such as performing miracles to cure the sick, feeding the crowds, or calling disciples to himself—and further, resurrection from the dead is deliberately eschewed by the Gnostic Jesus. In general, the historical presence of the person of Jesus and the realism of the common life of the first century that is the constant background of the canonical Gospel accounts disappears.[6]

Intrinsically, the difference between the Gnostic Gospels and the accepted traditional ones is not that the canonical four (Matthew, Mark, Luke, and John) were declared inspired by a fifth-century council of bishops at Nicea but that in them, Jesus appears as a far more interesting and important character. *He does things*—like curing lepers, giving sight to blind, feeding the four or five thousand, and, not least, dying and being resurrected. The Gnostic Jesus is about as interesting and effective as a typical middle-level manager (the author speaking from experience) but therefore is no longer a challenging figure to less-than-self-assured middle-class Americans.

The Gnostic perspective applied to the early Church tends to reduce the person of Jesus to "nice guy" from his appointed eminence as god-man; and, in this way, Gnosticism itself is changed from what it was in the third century—a daring refusal to credit the revealed Word of God, which was so contrary to accepted social norms that the movement had to find its expression in underground meetings, déclassé semi-religious groups, and a distinct theology that finds God not in the unconditioned "other" but within ourselves. In the current version of Gnosticism, the hitherto-daring refusal to credit revelation and the sense of the hidden truths known only to the *illuminati* is no longer a secret but announced from the housetops. But what once was a heresy and an abomination in the sight of hitherto Christian society has now become not a revolutionary declaration whose advocates face persecution and death, but a reassurance that faith in the One God does not really need to cost a person anything. Contemporary Gnosticism implies that religious seekers can indeed "have it all," including the love of God, the forgiveness of sin, redemption, and a sense of divinely assured approval, without the painful necessity of recognizing one's own guilt, trying to live up to the commandments, and fulfilling the requirements of the Abrahamic faiths of personal repentance, humility, and daily observances.

The doctrine of Gnosticism has been transformed, in the American vernacular, from a challenge to the worldview of revealed religion to a reassurance that middle-class lifestyles will be able to combine hard-edged theological positions that deny the existence of a personal God and refuse to acknowledge his commands with a devotion to personal comfort and a clear conscience. The main effects of contemporary Gnosticism are its denigration of revelation and the cozy assurance that unrefined human reason and common human experience are fully adequate to recognizing and understanding the nature of divine reality. The nature of divinity has thus been reduced to accommodate the limits of human reason in such a way as to fit the bourgeois desire for psychological comfort: an anodyne to guilt, anxiety, and persistent feelings of inadequacy. At the same time, Gnosticism provides a basis for contemporary religious seeking beyond the confines of formal religious organization, while also foreclosing the possibility of meeting with unconditioned divine reality. At one end of contemporary religious thought, Gnosticism asserts the essential relevance of religious belief based on the idea that religious truth does not escape human experience or transcend the bounds of human ability to think or imagine. At the other end of the religious spectrum lies the denial, by evolutionary atheism, that religion is of any rational use to mankind at all.

Evolutionary Atheism: Dawkins on the Nonexistence of God

I have written elsewhere of the phenomenon of physicists talking about God, which has taken place because the revolutionary developments in physical theory in the twentieth century have pushed the concepts of physics beyond the positivistic limits of nineteenth-century materialism and determinism.[7] Biology, however, is a different story, as its great modern theoretical development is evolution, which posits an entirely mechanical explanation of the origin of life-forms, a theory that has been so successful, it is alleged, that all fields of biological science are conformed to it; that it not only provides the essentially complete explanation of the origin of species but of all aspects of all living organisms, including their social behavior—and including that of the *human* species. Natural selection is proclaimed as the central doctrine of evolutionary theory, announced by one of its foremost advocates as "Darwin's dangerous idea"; it is promoted as the fundamental principle by which virtually all the significant aspects of human existence—including history, economics, the rise and fall of empires, the growth of civilizations, the personal relations among family members, ethical systems, and, by necessity, religion—are comprehensively explained. In the hands of advocates such as Robert Wright or Daniel Dennett, there is virtually nothing of consequence that the mechanical principle of natural selection cannot explain, their uncritical belief supported by the statements of evolutionary scientists themselves, some of whom even allege that natural selection can explain the origin of the universe as well as the origin of life.[8] In this manner of thinking, evolution leads to the conclusion that there is no God—or that if there is, his existence is irrelevant to our understanding of life.

Based on the popularity of his many books and his general fame in America as well as Britain, the foremost proponent of evolutionary doctrine and scientific atheism is Richard Dawkins. There is much to be said about Dawkins—about his tremendous appeal; his classically good writing style, which is both passionate and clear; his effective use of metaphor; his acquaintance with the intricacies of genetic and evolutionary science, which he explains to a general audience with conviction and mastery; his knowledge of historical and cultural facts, which he deploys throughout his presentations; and, not least, his air of unflappable confidence, even where his arguments are weakest and his knowledge insufficient. Dawkins began his public career as a notable science writer, writing books and essays for an educated audience anxious for knowledge about current science, but then he evolved from an outstanding science writer to a proseletyzer for a precise and dogmatic atheism. Dawkins has a strike-force mentality against religious belief, particularly the Christian faith, explicitly disdaining agnosticism as a temporizing position;[9] he vigorously attacks religion, which he perceives not only as an enemy to free thought but, in the words of another atheist scientist, Weinberg, as "an insult to human dignity." His attitude is unlike that of Bertrand Russell, who participated in a famous BBC debate with Fr. Copleston, the Jesuit historian of philosophy, about the existence of God. By contrast, Dawkins's approach is peremptory; he dismisses Aquinas's five proofs in three pages and Pascal's Wager in two. [10] However, such a brief treatment of two of Christianity's foremost philosophers is a sign not of the depth of his knowledge or his intellectual power, but of Dawkins's intellectual triviality as a philosopher of religion. As Dawkins deals with a succession of related topics in his book, *The God Delusion*, he tends to rely on breezy summarization and dismissive expressions to rebut complex and well-supported arguments with which he disagrees.

It seems impossible to imagine Dawkins arguing calmly with a religious advocate, willing to patiently pursue all the logical intricacies inevitable in discussions about divine reality and human understanding. However, the impossibility reflects the fact that Dawkins's manner of presentation avoids the larger intellectual implications of his own position while sweeping along in a rush of assertions and disparaging remarks about religious belief. Thus, apart from his account of biological history, which is limited to his own involvement in the late-twentieth-century developments in social biology, he imparts no general knowledge about the transition from teleological to mechanical thinking as a result of Darwin's evolutionary theory (much less in the history of science as a whole). And even though his own area of expertise is in the application of genetics to social behavior, he gives no general account of the organization of human society. The ontological status of genes as a halfway house between cells and DNA molecules, the reliance of genetic science upon the indeterministic mathematics of probability, and the difficulty of discovering the exact chemical pathways from specific genes to the exact somatic qualities of organisms (not to mention their behavior) present no problem for Dawkins, as he merely asserts (in *The Selfish Gene*) that the primary entity of biological reality is not the individual organism but the gene.[11] Finally, Dawkins's arguments rely on the belief that scientific rationality is the exclusively correct manner of human thinking and of explanation, a belief that he never defends or explains in detail. All in all, Dawkins gives the impression of a fast-talking advocate of a slightly disreputable position—like the flat tax or an all-purpose cleaning agent—not that of an engaged scientist explaining the resonances of his field in serious intellectual terms for the general public.

Readers looking into Dawkins's *The God Delusion*, seeking there an intellectual argument, will be disappointed, and serious religious scholars who reviewed the book were surprised about how little Dawkins knows about religion.[12] But such a reaction is somewhat unfair, since Dawkins's book is not intended to be a comprehensive argument or series of arguments about the nature of the Abrahamic God—or how humans ever came to believe in such a being, if its putative nature it as illogical as Dawkins and other contemporary atheists claim. Rather, the book has the tone of the self-help books that popular writers produce for a general readership, assuring them that they are "okay"—that is, not as insufficient as their mothers, spouses, bosses, friends, or society in general make them feel. So also, Dawkins's book is really intended as a self-help book for atheists, assuring them that, despite the overwhelming presence of religious belief in society, there are many people just like them, and that there are really good reasons for them to think well of themselves, and that their sense of alienation from society is somebody else's fault. (Dawkins gives lists of mutual-support organizations for atheists.[13])

There is, however, one significant argument in the book about the existence of God, and, as usual with Dawkins, it concerns a metaphor that was originated by the late astronomer royal of England, Fred Hoyle. It is well known and relies on the striking image of a windstorm sweeping through a junkyard and producing a jet airplane (a Boeing 747, to be exact). The vast statistical improbability of such an event characterizes the virtual impossibility of random events in nature producing life from inorganic materials, an event usually associated with evolution but in fact inexplicable by the process of natural selection, and also the impossibility of random mutation and natural selection producing so complex an entity as the human species. It is, Dawkins admits, a formidable argument, an up-to-date version of the argument from design, but he asserts that the logic of the image can be turned on its head so as to provide a "statistical demonstration that God almost certainly does not exist."[14]

In order to successfully complete his argument that complexity proves that God does not exist, however, Dawkins must provide not the incomplete and overly self-assured commentary he provides but an alternative explanation, one that can explain all the facts of complexity.

Dawkins replaces the creative power of God with natural selection, which he elevates from a theory about the origin of biological species to the ultimate explanatory principle that encompasses all aspects of known reality. To make his case that this biological principle of explanation can replace the creative power of the revealed God, Dawkins must use all his formidable ability as a writer in full command of the armamentarium of rhetorical devices. But the fact is that natural selection, when extended from a *biological* to what can properly be termed a "metaphysical" principle, does not have sufficient power to make the case for evolutionary atheism. In fact, it is a fair question whether natural selection can explain the actual origin of individual species, for evolutionary biologists must supplement natural selection with additional sub-theories, such as that species evolve in geographic isolation, in small populations, and by means of sexual, as well as natural, selection. But even granting that natural selection can fully explain the origin of species does not imply that it can explain the origins or the ultimate meaning of human behavior, despite the claims of social biologists who must argue their case against left-wing critics who make a strong case for a cultural explanation for human behavior (see Chapter 9).[15]

Despite any intellectual weaknesses in Dawkins's position, atheism has attained a social presence in the English-speaking world, as there are atheist websites, books, and organizations that promote their point of view. One visible activity is hiring billboards that snarkily contradict biblical belief (e.g., denying the divinity of Christ, or even just his existence, around Christmas time). In this way, atheism has attained a role as a "religious" option in America, although in an obviously contradictory fashion, it is based on the denial of religious belief. As a religious replacement, however, atheism of any sort—but especially one based on evolutionary reductionism—cannot provide an effective substitute, since atheism lacks the means to sustain human life emotionally or provide insights into how we ought to live our lives; except that is, in opposition to religious belief generally and monotheistic religious belief in particular.

As it happens, the bulk of self-identified atheists are not especially noteworthy for their intellectual ability (unlike Dawkins) any more than for their understanding of religion (very much like Dawkins). A final point can be made here: namely, that Dawkins's understanding of the arguments for God's existence assumes that God merely performs the function of an explanatory device, a function that, as a rationalistically inclined biologist, he finds grossly ineffective. But rationalistic arguments do not often inhibit religious believers from practicing their faith or believing in God or in the whole resultant panoply of biblical inerrancy, angels and saints, miracles, and life after death. Hence, the atheistic religious option, while now having attained a real social presence in American culture, remains a mostly oppositive reality.

Is God Reasonable? Explicating Benedict's Regensburg Address

Thus far, in this presentation of "fragments" of religious belief in contemporary America, we have described two limits that define two opposing ideas: that on one hand, we can discern the reality of religion in a spiritual manner to discover the God within, and that on the other, religion can be fully explained by evolutionary science, which reduces religious insight to a biological urge. The third fragment is the remains of the American religion itself, that is, the reassertion of monotheism, understood as that primary element common to the so-called Abrahamic faiths: Judaism, Christianity, and Islam. All three religions find their dogmatic origin in the figure of Jehovah in the Old Testament, with variations (Jesus as the son of God, Muhammad as God's last prophet). In choosing Pope Benedict XVI as its spokesperson, we are not attempting a backdoor method of Catholic apologetics, as if it were a defense of papal infallibility. Rather, Benedict's Regensburg speech is used here because it, more than any other recent event or document, brings out the monotheistic option in contemporary terms.

This speech, which provoked riots and vitriolic criticism from across the Muslim world, is mostly remembered as simply a news event—that the pope inadvertently, or perhaps deviously, attacked Islam as a violent religion and was answered by a response that made his point (at least one nun was assassinated in the aftermath). All of this is true, as far as it goes, for the Regensburg speech was an attempt by Benedict to explain to what extent such a concept as "that God is a Person," as revealed in the scriptures, is related to the standards of human reason—a significant point, since the traditional revealed religions of Judaism, Christianity, and Islam still do, demographically, at least, dominate the religious landscape of the United States and the world. Thus, an examination of the Regensburg speech is useful for the purposes of examining the contemporary religious scene.

The pope, in his own right a major theologian of traditionalist bent and by actual occupation a college professor, was revisiting the place from which he had started his scholarly career. He had been a professor at Regensburg before ascending the ecclesiastical ladder to reach the papacy itself, and the university had been the place from which he had started. Thus, Benedict's speech took the form of an address to university professors, which, in essence, the man Ratzinger really is. As such, the address had some interesting features, mainly because it addressed the issue of how reason approaches the reality of God. Because he was addressing university professors, Benedict apparently did not feel the need to explain himself in terms a layperson could readily understand—even a well-informed or well-educated one—but in terms fit basically for his audience of advanced scholarly specialists. As a result, it is a very compressed speech in which points are made in summary fashion, one after another, in a complex relationship but with all the points leading up to a final, explicit two-part conclusion. Fully explicating it in all its historical and intellectual ramifications would take a college course of one or two semesters, and at least one book has appeared to explain the speech[16]. Nonetheless, we will attempt a compressed explication here; Benedict's address is so complicated and yet so compressed itself that even a summary explication must be compressed. The explication is worth the effort, however, for the light it shines from a significant source on contemporary views of the relation between faith and reason, and because Benedict's speech maintained, at high visibility and in a fashion sensitive to contemporary concerns, the doctrine of revealed monotheism.

The Regensburg Address can be divided up into separate sections, starting with an introduction. Following a traditional approach in the giving of speeches, Benedict begins with a joke, and, again following tradition, it is an *old* joke. Because of an arrangement typical of German university organization made subsequent to the Reformation, there is both a Protestant and a Catholic faculty of theology at Regensburg; thus the joke once told by a skeptical professor: that the university had two faculties devoted to something that did not exist, namely God. But Benedict has a serious point as well, that at the university—despite the differences of opinions and the separation brought about by the specialization common to university faculties—the experience of the faculty at Regensburg was nonetheless that of making up a whole, of "working in everything on the basis of a single rationality." It is worth noting how contrary to certain recent trends in scholarly opinion this characterization of rationality is, opposed to deconstructionist and postmodern critical schools associated with Derrida, Foucault, and Rorty, which deny that truth exists as a univocal thing. Benedict had declared that the denial of the reality of truth is the greatest challenge facing contemporary culture, and John Paul II wrote an encyclical on the topic of truth.

From here, Benedict jumps into the substance of his address, taking on the topic of reason and faith by using the now-famous comments of an obscure late-fourteenth-century Byzantine emperor to make his first major point. The emperor, Manual II Paleologus, was head of the Byzantine empire, which had its capital at the city of Constantinople, then under siege by the Muslim Seljuk Turkish army. This was the historical setting of the debate in which the emperor, in dialogue with a representative of the Muslim Turks, decried, with some vehemence, the Islamic doctrine of holy war. "Show me just what Muhammad brought that was new, and there you will find things only evil and inhuman, such as his command to spread by the sword the faith he preached." This was the quotation that was given prominence by media reports and which, when made public to the world at large, initiated a series of riots and demonstrations throughout the Muslim world. Why Benedict used it provokes a question: why use *this* quote when a number of others were available for the point the pope was making about the relevance of reason in Islamic theology and theology generally? In fact, the reference seems gratuitous, unless this was the way in which Benedict was making an indictment of the more recent expressions of the doctrine of holy war.

The emperor went on to say, in a sentence given special prominence, "God is not pleased by blood, and not acting reasonably is contrary to God's nature." Benedict then goes on to point out that "The decisive statement in this argument against violent conversion is this: not to act in accordance to reason is contrary to God's nature." But while this statement is self-evident to the Greek emperor, since it depends on the tradition of Greek philosophy—extending from Xenophanes through Plato and Aristotle and St. John, who begins his Gospel account of Jesus's mission with the statement "In the beginning was the *Logos*"—it is contrary to the basic Islamic understanding of the nature of God, called "Allah." It is this disagreement about whether human rationality can in any manner, partial or elevated, be attributed to God that is at the heart of the conflict between Islam and the Christian religion; since, as Benedict points out, "for Muslim teaching, God is absolutely transcendent. His will is not bound up with any of our categories, even that of rationality." Benedict cites as a source about the transcendence of the will of God the important Islamic thinker Ibn Hazn, but he might also have cited the important and influential Islamic theologian Al Ghazali, who denied the relevance of Greek philosophy to the Muslim religion and thus left God alone in his majesty, his mere word alone the law for humankind, without recourse.

Next, Benedict proceeds to expatiate on the nature of Greek thought and its intimate connection to religious revelation, for in his view, the tradition of Greek thought was providentially placed in human history as the required means of understanding and, in effect, assisting the revelation's penetration into human souls and human history. There was, Benedict states, a "profound encounter of faith and reason" accomplished by the "rapprochement" between the tradition of Greek thought and the revelations of the Bible, including the Old Testament. He adverts to the ancient Greek translations of the Old Testament, the Septuagint, and makes a comparison between the self-revelation of the Lord to Moses in the burning bush and Socrates's attempt to "transcend myth." "A profound encounter between faith and reason is taking place here, an encounter between genuine enlightenment and religion." However, Benedict goes on to say, developments in later medieval Christian thought attributed to the nature of God a voluntarism, which ultimately could lead to "the image of a capricious God, who is not even bound to truth and goodness." Nonetheless, it was the convergence of biblical revelation, Greek philosophy, and Roman organization that created "and remains the foundation of what can rightly be called Europe." The subject of the unity of European culture and its grounds in Christianity is a subject of importance to Benedict and one he has pronounced on elsewhere.[17]

Despite the intimate and productive connection between biblical revelation and Greek thought, there have been, since the rise of the modern age, attempts at what Benedict calls "dehellenization." Indeed, he says, there have been three stages of dehellenization, and within the compressed address, Benedict spends more words on this subject than on any other, critically analyzing each of the three stages. The first stage occurred as a result of the Reformation during the sixteenth century, when the Reformers appealed to the Bible alone as the source of Christian theology and practice, dismissing the great medieval structures of thought that integrated biblical revelation with Greek philosophy, because, it was claimed, they covered over the true spirit of the Gospels. The second stage occurred during the nineteenth and twentieth centuries and is associated with the thinking of the liberal theologian Adolf von Harnack, who attempted to bring Christianity into harmony with modern reason, eliminating, as a result, such doctrines as Christ's divinity and the Trinity. Liberal theology of the time left the figure of Jesus as a great moral teacher who freed religion from its accretions of theological doctrine and Greek thought, but this made theology a historical and scientific enterprise rather than deliberations on divine revelation—in order, Benedict claims, to fit religious thought into modern university organization.

The third stage of dehellenization, which Benedict says is *now* in progress, is based on our experience with cultural pluralism; it objects to the synthesis of Hellenic thought and Christianity as a "preliminary inculturation which ought not to be binding on other cultures." Benedict responds that "the fundamental decisions made about the relationship between faith and the use of human reason are part of the faith itself," defending not only the Hellenic element in Christian theology but the need for the application of human reason *generally* in considering the things of faith. However, Benedict does not proceed directly from the second stage of dehellenization to the third and current stage but in between them interposes a critical analysis of scientific thought. He had said that von Harnack's liberal theology was both historical and scientific, but he is mainly concerned with its scientific aspect.

These passages on the scientific method make it clear that Benedict thinks that scientific reason of a mechanical sort has come to illicitly characterize all reason, with the result that theology and the academic fields that study humanity attempt to conform to it. From this desire for scientific legitimacy, bad consequences follow on an intellectual level and a practical level. Benedict begins by making an unusual but pertinent point when he describes science as a "synthesis" between empiricism, the experimental element "confirmed by the success of technology," and a Platonic or Cartesian element that "presupposes the mathematical structure of matter." The application of the scientific standpoint, however, forces the human sciences, such as psychology, sociology, and philosophy, to attempt to conform "to the canon of scientificity" which thus excludes the question of the nature of God as unscientific or "pre-scientific." Further, any attempt to maintain theology's claim to be scientific reduced Christianity to a fragment of its former self, but Benedict follows up by asserting, in effect, that if the consequences of the misapplication of scientific canons are harmful to our concept of God, the consequences are even worse for the concept of human nature. If science "as a whole is this and this alone," then man himself becomes reduced, as religious and ethical matters have no place within the collective rational purview defined by science, and the specifically human questions about our origin and destiny are relegated to the realm of the subjective. "This is a dangerous state of affairs for humanity," since attempts to construct an ethic from evolution, psychology, or sociology are "simply inadequate."

Benedict concludes his passages on scientific method by asserting that he has no desire to put the clock back but that rather he accepts the "insights of the modern age." But while "we rejoice in the new possibilities open to humanity, we also see the dangers arising from these possibilities." Benedict declares his intent of broadening our concept of reason beyond the scientific and, to make his final point about a dialogue of cultures, will appeal to the Platonic element, which he earlier stated was an essential part of the scientific standpoint.

In the final two passages, the intense compression of Benedict's address obscures somewhat his final point, which makes the case for a double civilizational dialogue. First is the need for a dialogue, within Western culture, of reason and faith, which is intended for "broadening our concept of reason and faith" beyond the limits of positivistic reason. This in turn will enable us to proceed to the second dialogue, "the genuine dialog of cultures and religions so urgently needed today," which is, to say, a dialogue among all the cultures of the contemporary world; most especially, although Benedict does not explicitly say so, he is concerned with a dialogue between the West and Islam. In other places, Benedict has called for Islam to renounce terror and violence as a means of defense of their religion and for "reciprocity" between Western nations, which permit freedom of worship for all religions, and Islamic nations, which do not.

Benedict's basic theme in his address is stated in his final paragraph: that the essential nature of reason itself is put in doubt in contemporary civilization, with deleterious consequences. Thus, reason based on a misunderstanding of how science operates precludes a recognition of the reality of God and the spiritual nature of mankind, while a refusal to recognize, in a theological framework, the rationality of God precludes a recognition that violence is contrary to God's nature. The first reference is to modern science and the second to Islam. Benedict makes two other points along the way: that dialogue between the West and other cultures is not likely to succeed as long as the West excludes the divine from the "universality of reason," and that the West is endangered because of its "aversion to questions which underlie its rationality"; and here, Benedict seems to be referring to the postmodernist denial of the possibility of truth. Benedict asserts, interestingly and contrary to the common belief that religious belief is contrary to reason, that "theology grounded in Biblical faith" is inherently reasonable because it engages "the whole breadth of reason." In effect, Benedict asserts that the only current cultural source that promotes a confidence in reason, as such, is revealed religion—but particularly, the Christian religion. "I know that people charge the Church with lowering reason, but it is just the other way. Alone on earth, the Church makes reason really supreme. Alone on earth, the Church affirms that God himself is bound by reason," Benedict might as well say along with G. K. Chesterton's detective priest, Fr. Brown.[18] In Benedict's theology, reason, the *logos* represented in the person of Christ, penetrates the whole of universal reality, including God and including "the rational structure of matter." Benedict's penultimate point is made with another quote from the Byzantine emperor that he quoted initially: "Not to act reasonably, not to act with *logos*, is contrary to the nature of God." Benedict finishes, finally, by saying that the rediscovery of the universal status of reason, which he refers to here as "the great *logos*," "is the great task of the university." At this final point in his presentation, however, it seems as if Benedict has too much confidence in the university as a setting where contending intellectual factions and differing points of view can reason together, especially with the scientific fields often, as Benedict suspects, promoting a narrow, positivistic view of reason.

Looking at the Fragments

In looking over these three contemporary takes on religious faith, we can quickly summarize the situation as a response to decline. The Gnostic view is pleasing and conformable to American individualism as well as middle-class desires for a materially and psychologically fulfilling life. But it severely limits the possibility of what the spiritual realm consists of—it is one thing to say that the journey to find God winds down within our very selves, assuming that God is a being separate from ourselves; it is quite another to say that the sense of the divine we find within ourselves is, in effect, all there is to spiritual reality. Gnosticism, while promoting the idea that a universal spirituality is available to all mankind, also puts a restrictive limit on what the spirituality may be. The Gnostic sensibility is well intentioned, and to its credit also, it refutes the blank materialism that both psychologically and doctrinally is offered with increasing prominence by a purportedly scientific point of view. Nonetheless, Gnosticism is just "too American," just as Bloom correctly identifies its current manifestation as a continuous theme in American thought, from Emerson and William James up to Elaine Pagels and Harold Bloom himself, with byroads into Mormonism, Adventism, and New Age philosophy. No doubt that Bloom is right that Gnosticism is a major constituent of American religious thought, but unlike the revelation made to the Israelites of the demanding and provident figure of Jehovah, whom Bloom has reductively explained as a creation of ancient rabbis,[19] the Gnostic revelation reinforces the laziest aspects of the American character by refusing to take us out of ourselves.

The heated rhetoric surrounding current evolutionary materialism makes it clear that it really is intended as a religion substitute, not merely as a reductive explanation.[20] It is not enough for its avatars, including Dawkins, Dennett, and E. O. Wilson, to show that religious belief provides a false image of reality, but to place what Wilson calls the evolutionary "myth" as the object of—if not devotion—at least *practical veneration* for the human race. Yes, they seem to argue, at the end there may be only the pit, and we die like dogs with no serious likelihood of our surviving the cessation of bodily activities, but isn't just the vision of the evolutionary epic itself so awesome that we really don't *need* anything else? We can, in this view, not be too concerned with the loss of the solace, support, and comfort provided by religion, although evolutionary writers do try to replace the religious sense of ethical reality with a naturalistic ethics. The universal search for immortality does not cease, however, even among agnostic and atheist scientists; only now, it is clumsily transferred from the religious to the scientific realm, so that hope in the risen body at the general resurrection, or in the transmigration of souls, can be replaced with cryonic freezing and speculation about the human race's outliving the planet earth by making our corporeal bodies into units of immortal information.[21] In the end, evolutionary materialism does not suffice as a religion substitute, even on its own terms.

Pope Benedict's address is a thoughtful re-presentation of the case for God, but it has a kind of double difficulty in which the elements combine to make up a contradiction. That is, on one hand, it is apparent that Benedict's case rests on a view of universal reason that is under attack these days from two separate sources: the school of postmodern literary studies that denies that expressions have meaning independent of social context, which thereby makes reason no longer univocal but divides it up into separate and unconnected pieces; and the scientific view that does promote a univocal idea of reason but which reduces it to expressions of physical cause and effect, denying that there is any spiritual component to reality. On the other hand, Benedict also appeals to a concept of God that appears, in a significant way, to go well beyond the limits of reason, for without question, Benedict's understanding of God as a Person rests on the doctrine of the Trinity. In his view, Christ is the *logos*—so designated in John's Gospel—who was present at the creation and is the Second person of the Holy Trinity; but the Trinity is a doctrine that must be based on faith, because in terms of the human intellect, the Trinity rests on a contradiction of three as one. Nonetheless, the apprehension that a statement can be true underlies the possibility of the universal kind of reason on which Benedict's case rests.

Notes

1. Donald M. Braxton, "Religion-and-Science Dialogue from the Vantage Point of Religious Studies." *Zygon, Journal of Religion and Science* (Vol. 42, No. 2), June 2007; 285–288.
2. Eric Voegelin, *The New Science of Politics*; Chicago, U. of Chicago Press, 1952; 107–132. Hans Jonas, *The Gnostic Religion* ; 1958.
3. Elaine Pagels, *The Gnostic Gospels*; New York, Random House—Vintage, 1989; xix.
4. Harold Bloom, *Omens of the Millenium*; New York, Putnam, 1996; 24.
5. ibid.; 25.
6. Pagels, op cit.; 26. H. Bloom, op cit.; 166.
7. John Caiazza, *The War of the Jesus and Darwin Fishes* ; New Brunswick, Transaction, 2007; 73–82.
8. Daniel Dennett, *Darwin's Dangerous Idea*; New York, Simon and Schuster, 1995; 177, 184-185.
9. Richard Dawkins, *The God Delusion;* New York, Houghton Mifflin, 2006; 46–54.
10. ibid., 77–79, 103–105.
11. Richard Dawkins, *The Selfish Gene*; Oxford, Oxford U. Press, 1989; 237 ff.
12. See the review by Terry Eagleton in *The London Review of Books*, (Vol. 28, No. 20), Oct. 19, 2006.

13. Dawkins, *God Delusion*; 375–379.
14. ibid., 113.
15. John Caiazza, "Political Dilemmas of Social Biology"; *The Political Science Reviewer* (Vol. XXXIV), 2005; 226–237.
16. James Schall, S.J., *The Regensburg Lecture*; St. Augustine Press, 2007.
17. Joseph Ratzinger and Marcello Pera, *Without Roots: The West, Christianity, Relativism, Islam*; New York, Basic Books, 2006.
18. G.K. Chesterton, "The Blue Cross" in *The Innocence of Father Brown*; London, Cassell and Co., Ltd., 1911.
19. Harold Bloom, *The Book of J*; New York, Grove Press, 1990.
20. John Caiazza, "The Religion/Evolution Controversy; How Two Mystiques Devolved into Politics" *Modern Age*, Spring 2005 (Vol. 47, No. 2); 104-112.
21. Barrow and Tipler, *Anthropic Principle*; 658 ff.

Chapter 13

Decline of the American Philosophy: A Tragedy in Three Acts

Given the failures of both science and religion to provide a sense of cultural unity and, as well, an overall basis for government policy that is accepted by a working majority of America citizens, where *then* to turn? That is, looking at things as they stand now, with what can contemporary society fill up the large cultural hole left by the diminishment of the authority of modern science and the decline of the American religion? It might seem as though philosophy might provide first the bridge between religion and science but also, and as important, a means of understanding both in relation to the American ethos. And in fact, America has created its own brand of philosophy, Pragmatism, a bona fide philosophic school. But to what avail? The sad answer to that pertinent question is the subject of this chapter.

Two Contrasting Views of the American Philosophy

America has produced its own philosophic school, created in the middle of the nineteenth century during a time of great expansion of national power, technological development, scientific investigation, and religious fervor, and which is still influential—namely, Pragmatism. The school has been the subject of two recent books, one by a well-known writer on cultural issues, Louis Menand, and the other by a highly regarded professor of philosophy, Nicholas Rescher. Both acknowledge the importance and influence of Pragmatism as a unique American school but are at sharp odds over its impact and the significance of its history.

Menand's book is called *The Metaphysical Club*, which refers to an informal discussion group of thinkers and writers in Cambridge and Boston, once used by Charles Sanders Peirce, who is the primary intellectual figure of the Pragmatic school.[1] Menand's book is comprehensive and detailed, yet it is so well written so that one can read twenty pages at a time without realizing it. No figure connected with the development of Pragmatism goes unnoticed, including Chauncey Wright, Benjamin Peirce (Charles's father), and Oliver Wendell Holmes Jr.; however, the main figures are the younger Peirce, William James, and John Dewey (although Menand does include Holmes as a major figure in the school). Menand's method is to write fairly comprehensive accounts of each of the major figures, while also putting Pragmatism in the context of American history, from the middle of the nineteenth century up to the twenty-first. He believes that the value of Pragmatism—which accounts for its "fit" in the context of American life—is that it proceeds from the notion that our ideas can never be proved true in an absolute sense but that, in effect, our beliefs are bets on the future. "Though we may believe unreservedly in a certain set of truths, there is always the possibility that some other set of truths might be the case. . . . the moral justification for our actions comes from the tolerance we have shown to other ways of being in the world, other ways of considering the case. The alternative is force. Pragmatism was designed to make it harder for people to be driven to violence by their beliefs."[2]

For Menand, Pragmatism's virtue lies in its denial of absolute truth—that is to say, in its projection of epistemological and moral relativism. However, this interpretation of Pragmatic philosophy is at odds with the intent of the philosophers themselves, at least to a degree. Peirce never denied that truth was accessible to the human mind, and he relied on the medieval philosophic notion of "Realism" to explain the relation of human knowledge to the reality of general scientific laws. He thought that scientific research was a historical process that would come ever closer to describing reality, a view that physicists currently pursuing the Grand Unified Theory would find a comforting justification for their enterprise. Even James, while describing the "will to believe," came down to a metaphysical notion of monism to explain the integration of matter and spirit. But for Menand, Pragmatism's chief virtue lies in its alliance with Progressivism, that is, as the intellectual basis for progressive and liberal policies that have been injected into American life over the past century. This alliance between "toleration" (a term rarely if ever used by any of the Pragmatic philosophers themselves) and progressive politics is a welcome idea to Menand's audience of good-thinking, well-educated, and progressive elites, which accounts for the fulsomely positive reaction the book has received.

Nicholas Rescher's *Pragmatism: The Restoration of Its Scientific Roots* is not the first book that he has written on the subject; its full title, in any case, gives the general ideas of where Rescher stands in 2012 on the subject of the Pragmatic school's place in American thought—and what he asserts is its general decline.[3] Rescher has written, over a long lifetime of philosophical research, on a large variety of topics, and his *Pragmatism* reflects that. It is a book written for specialists and not at all written in an easygoing or smooth manner as is Menand's book, for it includes many examples of sharp philosophical argumentation amplified by charts, historical references, and numbered points. In further contrast, it does not have the easy optimism about Pragmatism's connection to the American scene and especially its current state, for Rescher's view is decidedly pessimistic. One of the reasons for Rescher's pessimism is that his view of the history of Pragmatism includes later and contemporary examples of the philosophic school, including Richard Rorty. That is, while Menand does not go beyond the normal trinity of Peirce, James, and Dewey, Rescher includes Rorty, of whom he says, "with Rorty, Pragmatism collapses into a freewheeling, speculative free-for-all."[4] In short, Rescher's account is one of Pragmatism's decline from a robust defense of objective knowledge based on the community of scientific knowers into a feckless subjectivism. (He does have praise, however, for Susan Haack, a contemporary philosopher who has defended the objectivity of knowledge and whom he characterizes as representing the "harder" and "tough-minded" sort of Peircean Pragmatism.[5])

Rescher distinguishes sharply between a "soft" and a "hard" version of Pragmatism, identifying James as the progenitor of the soft version and Peirce as the founder of the hard Pragmatism, the latter which, for Rescher, is the true form of the philosophy. He not only defends the Peircean version but devotes the entire book to the argument that Pragmatism should return to its Peircean or truly "scientific roots." He describes, early on, three steps to accomplish this return: first, one should construe "success" as epistemic, not as affective; second, epistemic success is neither that of an individual or a group, but it is a "person-indifferent or impersonal success in scientific matters"; third, the objects of Pragmatic assessment must be the methods used rather than individual conclusions reached.[6] Overall, Rescher's view is more complex and more realistic than Menand's, whose book, while including well-researched history, is simply too comfortable with the solipsistic ideal of the Left, in which social progress depends (in part) upon the decline of the influence of true believers.

A Tragedy in Three Acts

It is argued here that the story of American Pragmatism is one of outright tragedy in the philosophic sense—a story of decline. Further, if the history of Pragmatism reveals a major figure, it is not William James but the school's putative founder, Charles Sanders Peirce. The sense of tragedy resides in the perception that if America were to produce a philosophic account of the relations between science and religion, or of scientific determinism and moral idealism, or a coherent application of scientifically inspired ideals to American democracy, that Pragmatism would produce it; but, alas, it has not been so.

The tragedy proceeds in three acts: in the decay of the Pragmatic idea of meaning and the future of scientific knowledge, from Peirce to James; in the decline from an attempt to satisfy both the religious and scientific beliefs of humankind to a justification of a utilitarian social philosophy, from James to Dewey; and, in the last act, from a belief in progressive ideals justified by Pragmatic philosophy to a frank belief that truth is too elusive to be identified by any one person or any one belief system, from Dewey to Richard Rorty.

Act One

Peirce is not highly regarded by partisans of William James, and Peirce himself seems to have acknowledged this precedence. Peirce wrote philosophy in highly technical terms (he invented a category of reasoning called "abduction," a category of metaphysics called "thirdness," and a theory of semiotics), which, however, was deeply original and philosophically accomplished but was not understandable to the general public—not even to an educated public. Peirce said of himself that in comparison to James, he was a mere ball of twine; and he understood that James was the effective popularizer of Pragmatism, but that he had also given public credit to Peirce as the founder of the school. Peirce was grateful to James, whose personality was engaging and outgoing, while Peirce was notorious for his extramarital dalliances (which led to the divorce from his first wife) and his apparent instability (he never was able to hold an academic position on a long-term basis). However, Peirce's version of Pragmatism was significantly different from, and philosophically superior to, James's.

Basically, Pragmatism is a theory of meaning—that is, of determining the true meaning and definition of concepts used in human reasoning. The theory states that the true meaning is to be found in the effects of how that term or concept is used in practical terms—that is, as used in real life, whether by housewives, laborers, philosophers, or scientists. Peirce used the example of the concept of hardness, and he cited as the basis of its meaning a scale of hardness, the so-called scratch test used by mineralogists. A sample of iron scratches a sample of copper, but iron in turn is scratched by a diamond; thus, the iron is harder than the copper, and the diamond is harder than iron.[7] This practical means of determining meaning was used, taken up by James, whose examples were often not scientific, however, but which more often concerned practical aspects of human life, including ethical choice and religious belief. This amplification by James indicates how his version was more approachable because it was less technical; however, James had left something essential out of Peirce's version of Pragmatism, namely that for Peirce, the meaning of terms was not in the practical consequences themselves (e.g., the scratches made by a diamond on iron) but in our *concept* of the possible effects of the test—"consider what effects, which might conceivably have practical bearings, we conceive the object of our conception to have. Then, our conception of these effects is the whole of our conception of the object."[8] For James, however, the meaning of a concept was essentially utilitarian; that is, it was in the practical effects themselves.

In his *Varieties of Religious Experience*, James used the term "cash value" to indicate how he evaluated the practical effects of belief; it becomes apparent, on a careful reading, that James's use of the term is not symbolic or a colorful metaphor; in an actual sense, James did look to the most practical effects of religious belief in order to evaluate them, finding no "cash value," for example, in the religious life of Teresa of Avila, the mystical writer and reformer of the Spanish convent system.[9] Without the conceptualization of the effects, James's theory was a kind of systematic and subtle materialism, while Peirce's retained a metaphysical aspect that asserted that meaning was a compound, somehow, of practical application combined with an immaterial, general concept. On this subject, Peirce referred to the controversy that had occurred in philosophy in the late medieval period over the issue of whether general terms had a reality in themselves or were merely sound, as it was said, emanating from one's throat. The schools of controversy were the "Realists," who thought that concepts such as goodness or justice had meaning, versus the "Nominalists," who denied true meaning to any term except those that referred to individual items. Peirce had discovered the writings of John Duns Scotus in the Harvard College library, which dealt with the issue and which he found useful in understanding the nature of scientific laws; were such laws mere compendiums of experimental data, or did they designate a reality in themselves (e.g., Newton's laws of motion)?[10] When he detected the difference between his "realist" understanding of meaning and James's, Peirce argued in a friendly fashion with James about it and declared that henceforth he would call his philosophy "Pragmaticism," a term so ugly, Peirce commented, that nobody would want to adopt it.[11]

Another contrast is worth mentioning, however briefly. Peirce was involved in experimental science, taking part in expeditions of the United States Coast and Geodetic Survey, and was well versed in chemistry, mathematics, and the history of science. Looking at the overall growth of science, Peirce noted that experimental science progressed by putting forth theories and then testing and disregarding failed hypotheses, and he noted also that science advanced by making discoveries and inventing ever more complex theories. Peirce thus arrived at the conclusion that while science as a whole continued to discover truths about the physical universe, it never would be able to claim that it had attained absolute truth. Thus, the growth of scientific knowledge was accumulative in historical terms; but in mathematical terms, it was *asymptotic*, in the manner in which a curved line such as parabola moves ever closer to its terminus but never reaches it. The final culmination of human knowledge lies in a utopian point at the end of human history, Peirce implies.[12] Despite its futurism, Peirce's view of ultimate scientific truth emphasizes his *realism*; that, for him, human knowledge was real, since human activity could discover those concepts that reach a correct description of objective reality.[13]

In James's thought, Peirce's grandly speculative theory of knowledge and scientific progress was gruffly reduced to "the will to believe." James did major work in the development of scientific psychology in a fashion that was neither reductive nor materialistic, thus refusing to follow the trend of behavioral psychology invented by his contemporary Watson and later developed by Skinner (for example). Rather, the development of a scientific but non-reductive psychology concentrated James's mind on the manner of personal belief as it occurs within the individual person and not in an overall sense. In elaborating a "will to believe," James left unaddressed the question of whether the objects of belief are real or invented, and therefore, he left unaddressed the question of what the effect is if someone believes, however strongly, in an assertion that is objectively wrong. "A live hypothesis is one which appeals as a real possibility to him to whom it is proposed. . . . [D]eadness and liveness in an hypothesis are not intrinsic properties, but relations to the individual thinker. They are measured by his willingness to act. The maximum of liveliness in hypothesis means willingness to act irrevocably."[14] James thought that belief was an activity of the human personality that was discoverable and worthy in itself, regardless of its object, and tended to temporize over the question of whether there was an objective reality that matched a belief. Thus, James's metaphysical monism, by which he imputed the qualities of both matter and spirit into ultimate reality, is also a means of evading sharp questions about the existence of God and the validity of the Christian faith. Peirce, on the other hand, devised a new proof for the existence of God, albeit one that he himself termed "neglected."[15]

Despite the fact that critics such as Menand and A. N. Wilson have elevated James to the pinnacle of importance as representative of Pragmatism in comparison with Peirce, in the view presented here, James's philosophy appears to be a reduction in range from Peirce's Pragmatism, including James's lack of a precise metaphysics and in technical detail as well. The transition from Peirce to James is only the first declension, the first act, of what constitutes the tragedy of American Pragmatism. The next stage of decline is the transition from James to John Dewey, who was James's student at Harvard College.

Act Two

Dewey's influence has been profound, less because he was a philosopher who, like Kant, wrote influential works than because he was an educator and representative of the Progressive tradition in American life in education and politics. In the transition from James to Dewey, the Pragmatic movement has left the nineteenth century and moved into the twentieth; and in the twentieth century, ideologies developed in the nineteenth, including socialism, fascism, violence as a tool of politics, and ideological nationalism, were put into practice. One of those ideologies, more benign certainly than those others, is Progressivism, which is based on the belief that human activity and social realities are best understood by rational means combined with good will.

The subject of Dewey's seemingly irenic philosophy is that true philosophy goes beyond the understanding of fixed realities, for rational thought must accommodate itself to ever-changing social realities and not insist on fixed notion of what constitutes learning, social reality, or human nature. The true aim of Dewey's philosophy reflected the practical rather than the theoretical aim of science, which is to go beyond mere understanding to control of physical nature; in Dewey's case, this meant changing the structure of society. Dewey was an activist as well as a philosopher; among his activities was organizing a committee of well-known figures to investigate Stalin's charges against Trotsky. (The committee cleared Trotsky, a verdict that became moot after one of Stalin's agents killed him in Mexico.) Before he became a professor of philosophy at Columbia University in New York, Dewey had made his reputation by running and organizing an experimental school at the University of Chicago; and, in fact, Dewey is best remembered as an educational reformer and as an advocate of "progressive education," especially in childhood. The Progressive ideal of education promotes an exploratory and less-restrictive form of education, denigrating requirements of memorization, standardization, and strict discipline; as such, it has become controversial, since it is claimed to have led to a decline in educational standards.

Dewey was an acute thinker whose writings on philosophy are diffuse in style but which also indicate a wide range of knowledge. Both his style and his erudition are apparent in his book *Experience and Nature*, published in 1920. This long work of approximately 350 pages places Dewey's thought surprisingly in the tradition not of American Pragmatism but German Idealism—that is, not in the direct tradition of Peirce and James but in that of Kant and Hegel.[16] Despite the historicist and idealistic aspect of this presentation, however, in his later work, *Reconstruction in Philosophy* (1920), Dewey asserts that philosophy should no longer concern itself with a static view of reality in an unfruitful search for the transcendent reality, but rather that it should accommodate itself to the here and now of social reconstruction.[17] In this, Dewey resembled Marx, who famously asserted that the job of philosophy was not to understand the world but to change it, a sort of slowmotion version of Marx's revolutionary idea of social reconstruction.

In Dewey's Pragmatic vision, philosophy and social change go hand in hand; he argued, therefore, that philosophy should give up its unfruitful search for divine or transcendent reality of knowledge or morality and rather align itself with social change and the scientific understanding of nature. After all, he argued, so much had changed since the end of the First World War, and a reconstruction of philosophy was what was needed for the now-obviously required reconstruction of society.[18]

The practical effect of Dewey's intellectual and public efforts was to promote social change on the basis of a progressive ideology, which meant throwing aside millennial traditions, beliefs, habits, and seeming inefficiencies, less by detailed philosophical argument than by appeals to public awareness, promoting changes in the law, and encouraging policies enacted by the executive branches of governments and lesser bureaucracies. However, this general approach also meant foreshortening the range of philosophy by limiting to material, albeit social, realities. Dewey's is not a pure materialism, more a naturalism in which no transcendent horizon is understood to exist, and therefore no effort, intellectual or social, need be attempted to reach it. Indeed, such attempts, whether Plato's philosophy or biblical theology, obstruct the march of progress to a more socially efficient world. From James to Dewey, then, constitutes the second act of the tragedy, in which a Pragmatic philosophy that attempted to mediate between the ideal and the material orders has now transferred its allegiance entirely to the material order, albeit understood in terms of social reconstruction.

Act Three

The final act of the tragedy of the Pragmatic tradition in America is the decline from the philosophy and public stance of Dewey to that of Richard Rorty, as we pass from the early to the late stages of the twentieth century and into the twenty-first. This act is shorter than the first two, more an ironic commentary on what has happened before, for a reprise of the two prior acts of the tragedy here turns into a kind of bitter comedy. For Rorty, the attempt to base thought and action on a firm basis is foredoomed to frustration and self-contradiction. The attempt by Peirce and James to find such bases of thought and action—the action motivated by thought, the thought a result of the action that human agencies enact in the world around them—is given up by Rorty as he descends into the toxic mists of epistemological relativism very much resembling the influential postmodern critique emanating from European philosophers and literary theorists such as Stanley Fish.[19] Relativism in epistemology and ethical theory, combined with anti-essentialism, is a powerful meme operating in our times and appears to be the long inheritance of the two world wars that destroyed the traditional culture of Europe. The uneasy sense that there is no essence to things, no moral or physical center to which we can attach our being or find meaning, is reflected not only in the tomes of existential philosophers but in popular culture—for example, in the movie *The Discreet Charm of the Bourgioisie.*

Rorty made his reputation among philosophers with his book *The Mirror of Nature*, which is a severe criticism of Descartes as representative of the ideal of modern science that there is an objective truth to be known, and that modern science is the best way to know it.[20] No more reference to St. John's "You shall know the truth and the truth shall set you free"; rather, the prevailing sentiment portrayed in Rorty's book is that confidence that we can know the truth is misplaced, which leads to arrogance and oppressive behavior toward others. In Rorty's version of Pragmatism, we see the final decline from the positive assertion by Peirce that truth can be known over time, given the human race's determination to discover it, to the notion, passing through stages of decline in James's and Dewey's philosophies, of Rorty's denial that truth, even in scientific terms, can ever be known.

Despite Rorty's epistemological relativism, he retained his desire to assert the rightness of the same set of Progressive values and liberal policies promoted by Dewey. This contradiction left him in a quandary neatly expressed in the phrase "Liberal ironism," which, of course, is more the recommendation of a stance—a fashion statement, perhaps—rather than a serious intellectual or political position. This is in contrast to the defense of liberal values set out in John Rawls's *A Theory of Justice*, in which arguments from philosophy, politics, and economics were deployed in an intricate example of formidable scholarship—or Karl Popper's energetic defense of a Libertarian social philosophy based on a critical epistemology.[21] In contrast, Rorty satisfied himself with critical analyses of Derrida, Nietzsche, and Orwell instead of doing the constructive work of proposing and defending a theory of the Liberal state. Rorty states:

> The ironist by contrast is a nominalist and a historicist. She thinks nothing has an intrinsic nature, a real essence. So she

> thinks that the occurrence of a term like "just" or "scientific" or "rational" in the final vocabulary of the day is no reason to think that Socratic inquiry into the essence of justice or science or rationality will take one much beyond the language games of one's time.[22]

This fey attitude, distanced by a cool eye from the recurrent crises, political division, and cultural warfare in America in the present day, has an emphatically negative consequence. The inference that can be drawn is that the state may act without reference to the two elements that otherwise would act to inhibit it—namely religion, which probably never attracted Rorty's attention, and now science. Science, once seen as the avatar of rationality, is now seen as a social construction of reality erected for the empowerment of a class of experts and administrators and therefore not a means of discovering objective truth. Without recourse to either religion or scientific reason, state policies are now impelled by two major factors: by ideologies, whose influence is permitted and sympathetically understood, such as economic equality; and, most threateningly, by the drive for power in itself. The fuller consequences of this fact of the libidinal drive for power by the modern secular state are dealt with in Chapter 15.

Notes

1. Louis Menend, *The Metaphysical Club*; New York, Farrar, Straus and Giroux, 2001; 200.
2. ibid.; 440.
3. Nicholas Rescher, *Pragmatism: The Restoration of its Scientific Roots*; New Brunswick, Transaction, 2012. *Methodological Pragmatism*; Oxford, Blackwell, 1977.
4. ibid.; 10.
5. ibid.; 17, 18.
6. ibid.; 17.
7. Charles S. Peirce: *Selected Writings* (Philip P. Weiner ed.); New York, Dover, 1958;124–126, in the essay "How to Make Our Ideas Clear."
8. ibid.; 124.
9. William James, *The Varieties of Religious Experience*; New York, Modern Library, 1999; 379.
10. Peirce, *Writings*; 299–300.
11. ibid.; 181, 204.
12. ibid.; 138.
13. ibid.; 108, 9.
14. William James; "The Will to Believe" (Part I).
15. Peirce, *Writings*; 358–379.
16. John Dewey, *Experience and Nature* (2nd ed.); Lasalle, Open Court, 1961.

17. John Dewey, *Reconstruction in Philosophy* (enlarged ed.); Boston, Beacon Press, 1957; 24–27.
18. ibid.; v.
19. See, e.g., Stanley Fish, *The Trouble with Principle*; Cambridge, Harvard U. Press, 1999.
20. Richard Rorty, *Philosophy and the Mirror of Nature* (anniversary ed.); Princeton, Princeton U. Press, 2009 (originally published 1979).
21. John Rawls, *A Theory of Justice*; Cambridge, Harvard U. Press, 1971. Karl Popper, *The Open Society and Its Enemies*; New York, Harper, 1962.
22. Richard Rorty, *Contingency,Irony and Solidarity*; New York, Cambridge U. Press, 1989; 74, 75.

Part V

Unresolvable Differences

Chapter 14

The Arrival of Techno-Secularism

What has been attempted thus far in this essay is the importation of the religion-science controversy in it classical form into the context of contemporary American culture. But what appears up to this point is a discordant mixture of competing elements, for it seems that as a matter of cultural fact, neither religion nor science provides a basis of unity for American culture. But certain themes do emerge from this analysis: First, that science, as well as religion, has been overcome by the arrival of a level of technological dependence that has assumed its own force and existence (what is here termed "techno-secularism"). Another and truly disturbing element that emerges is the power of the secular state. Culturally, without the social force of either scientific rationality or of religious belief, the secular state does not feel the effect of an internal or external agency by which to measure its policies, leaving the secular state to express the force of its will without serious resistance or opposition. Finally, religious belief in a traditional form persists, even though diminished in its effectiveness, in a transformed manner. Its most effective expression is a general monotheism; is the former description of American religious culture as "Protestant, Catholic, Jew" now changing possibly to "Christian, Jew, Muslim?"

Origins of Techno-Secularism

Robert Coles claims in *The Secular Mind* that the origins of secularism and its recent upsurge are not to be found so much in scientific thought but in the nature of religious faith itself. As religious ideals rub up against the quotidian, secularism as a form of doubt becomes the inevitable psychological complement to faith, he argues. He quotes a conversation with Catholic activist Dorothy Day: "I think you underestimate doubt as a constant part of faith—in any century;

and I think you are making too much of science (and social science) as the (recent) 'causes' of secularism. I don't deny that today there is the authority of scientific knowledge to elicit or encourage or give a kind of imprimatur to secularism; but for Heaven's sake, the secular world has always been 'there' or 'here.'"[1]

On the other hand, Coles's meditations are in response to the dramatic event of Freudianism replacing religion in the understanding and treatment of individuals suffering from mental distress. As a psychiatrist and a man of religious sensibility, Cole might well be expected to put the issue of contemporary secularism in the context of the stresses attached to personal religious belief, yet the circumstance of Coles's writing his book belie his understanding of doubt as a constant twin of religious faith, since what he is describing is the displacement of religious concepts by those of science. It is true, as Dorothy Day points out, that there was secularism before there was science, but now secularism has become a social movement defended by philosophers, scientists, politicians, and writers. It is not enough, therefore, to see secularism as another name for doubt and as the inevitable complement of religious faith, for this subtle psycho-theological observation does not explain the roaring reality of rampant secularism seen in the present day, and of science in the form of technological application as its chief agent.

How science changed from a form of praise of God the Father's creation by such early giants as Galileo and Newton into an aggressive competitor of religious faith was a long process, but one influential American philosopher provides an illustration of how this transformation happened, and not as an unintended consequence but deliberately. In 1902, William James published his Gifford Lectures, *The Varieties of Religious Experience*. One hundred years later, the book has been republished by the Modern Library, which reflects its importance as a cultural event, for its initial publication was a transition point from a science whose purpose was to reflect the glory of God to a science whose intent is to replace religion with the glorification of the human intellect. *Varieties* has been influential just because it is not an example of blatant atheism but because it proceeds more subtly and more powerfully. James's book is a phenomenology of religious experience, examining religious belief not in doctrinal or historical terms but by means of the then newly developing science of empirical psychology. The book largely consists of reports of religious experiences, internal states that the subjects connect with divine or other external spiritual entities, which James analyzes in terms of his pragmatic theory of truth. James's conclusion is that such experiences do not validate any particular religious tradition, and especially not the Calvinist Protestant tradition. James's case against the Protestantism of his day as a form of psychological strain and excess is easy to make, since he defines all religious experiences in psychopathological terms and applies a pragmatic, practical, businesslike criterion of meaning to them.[2] On the other hand, James's phenomenology of religious experience tends against a reductive point of view, for he takes reports of religious experience at face value and thereby eliminates the intellectual possibility of scientific materialism. Metaphysically, James reached the conclusion that the variety of religious experience was best explained not by seeing reality as a duality of mind and matter (i.e., as a competition of religion versus secularism) but by seeing reality as a monism that combines both elements of mind and matter and which could in effect support either religion or secularism. Such an approach may seem expansive or contradictory or even two-faced, leaving James's readers to wonder which side he is really on. But as a practical matter, given a choice between the opportunity to make good in a time of burgeoning industry and commerce, or the stringent demands of Calvinist ethics, who is likely to embrace the latter?

Socially, what James accomplished was to provide a scientific rationale for displacing evangelical Protestantism with a variety of free thought among the elites of American society. His philosophy of religion made possible for them indulgence in new kinds of religious experiences, including spiritualism, séances (in which James participated), reincarnation, theosophy, and Eastern mysticism, without the traditional Christian elements of judgment and hellfire. In this way, James was the prophet of current, self-affirming "new age" religion. It was a time when technology and industry were transforming American life, and the glorification of business and greed seen in the gospel of Herbert Spencer's philosophical evolutionism was destined to come into conflict with the rigorous ethics of the Calvinist Christian gospel, which counseled humility, doing good for others, and subservience of self, which are hardly the ethical ideals to suit the exploitation of business opportunities. James greases the ways for a secular outlook by applying the pragmatic theory of meaning to religious beliefs in which their practical effects are their warrant for validity and value (see Chapter 13). Frequently in *Varieties*, James uses the phrase "cash value" as a metaphor for the Pragmatic theory of meaning, but in time the reader begins to realize that "cash value" is not a metaphor but designates the real thing, the real sense of what James believes the value of meanings and beliefs to be.

I have presented James's *Varieties* as if it were an ideology—that is, less the product of the independent thought of a philosopher than a reflection of the change in the technological structures of production, to use a Marxist turn of phrase. This is justified because the practical effect of their writing and thought is the criterion James employs in judging others, and the cash value of his *Varieties* is that it gave leave to the elites of American society to disregard the stringent ethics of Calvinistic Protestantism and invent an ersatz spiritual life for their own comfort. In this way, the business ethic was able to overcome the Protestant ethic, and technology succeeded in displacing religion to give secularism a social reality it had never had before in American thought and life. By emphasizing the cash value of religion, James had, in effect, turned it into a technology (i.e., means of production); but as I shall point out, the technologization of our culture has had the same effect on science itself.

The Magical Quality of Techno-Secularism

In our day, technologically based secularism threatens to displace religion from the national consensus; however the success of secularism is based on the effects of technological advance rather than on the victory of scientific ideas in the conflict with religious beliefs. How this happened can be gleaned from a remark by science-fiction writer Arthur C. Clarke that "any sufficiently advanced technology is indistinguishable from magic." It would be impossible to describe the effect of technology on contemporary life in a few words, as its presence is ubiquitous and its manifestations manifold. One point of recent technological development is worth making in general terms, however; namely, how technology itself is evolving from its nineteenth-century mechanical phase to a twenty-first-century phase that Clarke called "magical." That the difference is qualitative and not merely quantitative can be imaged by means of the steam engine and the personal computer.

In the Victorian times, steam power was the main force used for technological advance, most obviously the steam train engines, which even today, have not lost their evocative power. The point of the nostalgia is that every aspect of the technology of a steam engine was open and available for inspection: fire box, water pipes, smoke stack, steam valves, reciprocating rods, and driving wheels. The immense force of steam power pulling tons of iron and steel was understandable just by observation; there was romance but no mystery. Now, however, the most typical example of twenty-first-century technology is not the steam engine but the computer, whose product is not motion and force but organized information. The appearance of words on a computer screen can, of course, be connected to the keys we hit on a keyboard, but we cannot see the causal links from the keyboard to the screen, for the keystrokes are transmitted by an electronic and not a mechanical process—unlike the typewriter, in which the keys are connected to push rods, that have a type face for each letter, which strike the paper on the platen. Further, if we take apart the computer, we see elements such as an electric motor, fan, transformer, cables and wires, boards, and chips, but they do not directly convey how the process of computerization takes place since the calculations, sortings, and arrangings of data are done within the chip, which has no moving parts; and thus the computer's workings remain a mystery even after inspection of its innards. Effectively, it is a "black box" whose inner logic and workings are unknown to us. Unknowable, it is unrepairable by its users, and like many appliances today, repairmen will tell us that it is cheaper and more efficient to replace it than to fix it. We have retreated from mechanical explanations in terms of Victorian forces—explicit, competent, and muscular—to a postmodern realm of magical effects whose causes cannot be explained—mysterious, astonishing, and the province of experts who regard us with contempt.

The transformation of technology from a mechanical to a magical phase indicates its enormously enhanced power and influence; steam trains provided a visible replacement for horses and walking, but information processing is so prevalent that even if we do not own a personal computer we are nonetheless beholden to computerization in stoves, television sets, weaponry, and libraries. Technology has become so prevalent in manipulating and transforming our life-worlds that it has, in a way, overcome theoretical science, for by "science" the general public now perceives not an empirical/mathematical explanation of physical phenomena but the power to change our lives, to make them more comfortable by making our personal environments more responsive to our wishes. According to one qualified observer, "Americans like science and technology but are happy enough not knowing very much about it . . . [w]e can blame the state of public education for this, but there is something willful in it."[3] Applying

William James's terminology, it has been said that technology is the "cash value" of science, and as in the case of religion, the reduction to cash value empties out the true value of the scientific enterprise, which is to increase mankind's knowledge.[4] In this way, science is reduced to its technological expression and the scientist perceived less as a discoverer than a magician, a *magus*. Our personal environments have become so much the result of technological manipulation that when we reflectively look upon them we perceive the creative power of human scientists, whereas before when we reflected upon nature we could see the creative power of God. Technological effects have acquired a life of their own, achieving a qualitative level of change so that now technology has, in effect, its own ethics, theology, and unanticipated consequences. The displacement of religion from our civic life is largely the effect of technological ubiquity and power rather than result of direct cultural and intellectual causes, a phenomenon that may be termed "techno-secularism."

One particularly important result of technological ubiquity is the degree to which it has sustained and extended the power of the state over our lives. The increase in the amount of data and reports required of corporations, colleges, businesses, and nonprofit institutions could not have taken place without new developments in technology. In turn, the increased sophistication of technology has empowered an exponential increase in the amount and particularity of regulations imposed upon individuals and institutions in civil society. Thus, the recent electronic revolution has only intensified the impulse to the bureaucratization of power that followed upon such technological advances as air-mail delivery, carbon paper, telephones, skyscrapers for office space, mechanically powered transportation, typewriters, and, not least, automatic weapons. Computerization in the ability to electronically replicate, organize, and transmit data over the Web have made possible an incredible expansion of federal and state control over our lives so that the technologically amplified power of the bureaucratic state has made the state the chief object of concern and worry of its citizens, since its permission and benefits are required to conduct virtually every aspect of the daily business of contemporary life.

The Ethics of Techno-Secularism

What we emphasize here are techno-secularism's ethical and religious dimensions that are mediated through its implicit concepts as well as its practical effects. Techno-secularism has an implicit ethical theory that is instrumental, accepting that what technology can provide should be used for the betterment of the human condition without consideration of prescriptive ethical rules and humane traditions. It is utilitarian, opting for the greatest good for the greatest number, only that the "good" is understood in relentlessly material terms, that is, terms amenable to technological control. The ethic is eudaimonian rather than hedonistic, concerned with bodily well being rather than the maximization of pleasure. Techno-secular ethics is diet conscious, encourages the drinking of light wines rather than beer or whiskey, is antismoking, promotes safe sex, and is mightily concerned with a long, fulfilled, healthful life. It is nonetheless a materialistic ethic with a "horizon" that ends with death and encourages a fearful rather than a heroic lifestyle, justifying abortion and euthanasia because of the demands that children and the aged make: disfiguring women's bodies, taking up precious time (the one commodity that oddly technology cannot provide in abundance), and stultifying the careers and personal goals of both men and women. Techno-secularism is fearful even before the point of death, fearing the incompetence and dependence of old age, while sequestering death, unseen, to hospital rooms and the ministration of experts on death and dying. Instead, emphasis is put upon extending the period of healthful, fulfilled living for as long as possible, putting forth the possibility of extending youthfulness by medical technology and of technologically sustained immortality in the form of cryogenics and cloning. Avoiding the inevitability of death, techno-secularism refuses to deal with the issue of what comes after death—if anything—and its ethics is formed without reference to God and religion, since they are possibilities that extend beyond its horizon.

With regard to religion, techno-secularism attempts to "empty out" the doctrinal teaching from religious belief in order to co-opt religion's ability to change lives and to generate major social movements, which are, in James's terms, religion's "cash value." Techno-secularism has a fear of religion's ability to authoritatively move people and social events and occasionally attempts to refocus religious belief from religious ends to those in line with the aims of the bureaucratic state. However, for religion to "work," the religious believer must actually think that the objects of his beliefs are real and that the doctrines of his religion are true. Techno-secularism hits the fatal shoals on this point, for it cannot provide a doctrine that it believes itself and yet that will motivate others in a religious way. Unable to divorce cause from effect—the content of religious belief from its effectiveness as a personal motivator and social force—techno-secularism relies on the smooth and unnoticed transition from faith-based explanations to scientific causes, as a result not of logical arguments but of the ubiquity of technology in our daily lives. In this way, magic has made a revival as the unseen scientific causes of technology are appealed to for the improvement of our lives, while true religion is trivialized and marginalized seemingly without effort.

Must magic prevail? Both modern science and revealed religion are united on the point that "magic" in the ancient sense, by which incantation and commerce with spirits could influence fate, is a superstition unworthy of acceptance by educated people; and in modern times, "magic" has come to describe a form of entertainment in which a magician performs tricks on stage and performs illusions. The doves do not appear magically from the wave of a colorful kerchief, since they were already in the magician's sleeve; the woman does not really float in the air, for she is suspended on thin wires hidden by the darkened stage. The magic that prevails in the dominance of technology in contemporary life is also a form of fakery, for its effects depend on the highly trained intelligence and hard sustained work of armies of scientists, researchers, technicians, and planners, all of whom work on materials already provided in nature.

The Infant Stem Cell Controversy

During the administration of President George W. Bush, there arose a fierce controversy that directly involved the elements under discussion in this chapter, namely the social influence, ethical complications, and political power of techno-secularism. The issue was stem cell research, a very promising new line of medical investigation, particularly involving *infant* stem cells. The controversy, now seemingly quiescent, is revelatory because it highlights the issue of how or whether religious doctrines should intrude into scientific research, especially when that research is presumed to be exceptionally promising and valuable according to the purposes of the state, since it gives the promise of great relief to terrible illnesses among its citizens. Stem cell research was promoted, frankly, as a near-panacea for such afflictions as Alzheimers, spinal-cord injury, and Parkinsons—that victims of spinal-cord injury would rise up from their wheelchairs was literally claimed by a political candidate, in an awkward copy of biblical miracle stories as when Jesus or Peter say to believing cripples, "Rise and walk." Only now, science would replace faith, and the miracle would be available to everybody regardless of race, nationality, creed, or gender through the practical effects of medical research and the ministrations of a benevolent state.[5]

The controversy was made instantly contentious when a panel of ethicians relied on by Bush claimed that infant stem cell research, since it destroyed human embryos that otherwise could be matured into live human beings, was an assault on the intrinsic dignity of human beings. The issue was not, as these contrarians made clear over and over again, stem cell research itself, but *infant* stem cell research, because it requires the destruction of human embryos, which, as they continually explained to the public, are in fact proto-human life (i.e., babies in process, as it were). The distinction did not carry much influence in the public debate since, as the advocates of infant stem cell research (ISCR) pointed out, infant stem cells are *pluripotent*, that is, able to be developed to match any organ of the human body; this fact was elevated to an almost magical property and was the basis of much overreach in the prediction of the medical successes that ISCR would bring about. Bush nonetheless did not give in and, impelled, as he said, by moral and religious considerations, and yet aware of the promise of such research, struck a compromise. Federal policy would allow the infusion of federal research grants for work on certain "lines" of stem cells already in existence, but not support any new lines; this policy did not restrict ISCR as such, but only the availability of federal monies for research. Bush was promoting no new law that would forbid infant stem cell research (although legislation has been passed forbidding human cloning). Bush's well-intentioned compromise allowed private funding to be used for ISCR as well as limited federal support, while giving serious consideration to the ethical objections.[6]

But Bush's compromise did not satisfy the proponents of ISCR, whose reaction was not merely critical but enraged, calling his administration "anti-science" and his policy an intrusion.[7] There was in their view, apparently, no reason for considering the ethical objections, since the medical researchers seemed to feel that their demands for research grant money was in fact an entitlement; Bush's critics were especially annoyed because the promise of ISCR was overwhelming in their view. In the end, the scientists and secularists got their way, as the Obama administration, which followed Bush's presidency, almost immediately lifted the restriction on federal funding for infant stem cell research. However, at that point, an ironic consequence had developed, for it now appears that although research into the use of stem cells for therapeutic purposes has been successful and that effective cures are being developed, that these results are almost exclusively coming from non-infant stem cells, from placental or birth-cord stem cells, or from a patient's *own* stem cells.[8] There have been few such practical successes from infant stem cells. This controversy has above all revelatory aspect, since it reveals the points at which the claims of modern science, unfulfilled in this case, are used to support the policies of an explicitly secular vision of the modern state. The controversy also points up how the actual results of scientific research—in this case, that infant stem cells were less useful than those from other sources—can simply be ignored in pursuit of a state policy.[9]

A closer look and analysis of the ethical objections is called for; what were the objections made against infant stem cell research on the ethical level? There are three, first of which is an application of the "slippery slope" argument. Here the point is that once it becomes permissible to create human embryos for the explicit purpose of using them for experimental research, then there is no inherent stopping point by which to argue against destroying fetuses that are further along in development. The embryo used for experiment in fact consists of only a few cells, highly organized it is true, but, the defenders argued, not in any actual way a human being. The problem for the defenders of ISCR is that there is no obvious point at which human fetal experiment becomes manifestly objectionable, up until actual parturition. Even then, some advocates of abortion rights have argued that the human infant, so heavily dependent on its parents and society in general for its support and existence, has not acquired even a right to life until further along in its development—at the point of language acquisition, for example. (Peter Singer, the prominent Utilitarian ethical philosopher, has made this argument.) Despite the lack of an intellectual stopping point on the slippery slope, it does seem as if by the middle of the fetal journey, say four or five months at the latest, that there is a general sense that experimentation is ethically forbidden—but this is only a general sense, that is, a sentiment; and sentiments are easily overridden at times of necessity. In any case, the defenders of ISCR, in replying to the slippery slope argument, claimed that it was effectively answered by legislation forbidding (not just denying federal funds for) human-cloning experimentation.

The second objection to ISCR is that it offends the *dignity* of human beings as such; this argument has been made most often by ethicist Leon Kass. ISCR, he argues, reduces human beings to mere examples of the mechanics of cell biology with no reference to their actual humanity.[10] Once in a petri dish, observed under a microscope, separated chemically gene by gene, taken apart cell by cell by sharp instruments, the complex network of human embryonic cells loses all relatedness to humanity, becoming a manipulatable phenomenon no more worthy of respect than any other microscopically observed set of cells. But respect for human dignity does not have the required sharpness of focus to be useful as a moral imperative, as a bright line beyond which we will perceive that a certain kind of action is immoral. If ISCR is immoral because it violates inherent human dignity, what about *in vitro* fertilization or birth control? Kass, the best-known defender of the concept, is consistent in denying the morality of *in vitro* fertilization and birth control. No proponent of ISCR is likely to deny the concept of human dignity itself, as B. F. Skinner did, but their argument in opposition may be that a claim based on the concept of dignity is too vague, and that arguments based upon it that prohibit ISCR are arbitrary, because they are not based on a clear argument leading from explicit premises.

A third counterargument against opposition to ISCR is more implied than stated, probably because it does not have to be made explicit in order to have an effect—that is, that one of the bases of opposition to ISCR is religious. It was apparently not felt necessary by his critics to say any more than to point out Bush's references, however passing and indeterminate to religious belief, for the obvious assumption is that religion has no place in politics at all, particularly in discussions about government policies. However, if religious ethics has any place at all in any possible debate over government policy, it would seem to be here in the discussion about the origins and value of human life itself. The exclusion of religious beliefs may have been taken for granted by supporters of ISCR, but it is on its face an indefensible exclusion. Science itself cannot give the definitions we need as a people, since medical science in this case is interested only in its ability to carry on medical research; for medical researchers, the issue was funding for a very promising line of research. That such research, or any research, might be immoral is of course important to scientists, but they are as likely as any group of imperfect human beings to see things in their own way, to be prejudiced toward their own interests, and so should not be allowed to be judges in their own case. In this instance, it seems that their zeal for pursuing research possibilities overcame any possible moral difficulty.

What then is the positive argument for ISCR? It is entirely Utilitarian, that whatever "sacrifice" of human dignity is made by what is, after all, at *best* a clump of highly organized cells that is proto- but not actually human, the positive effects are so overwhelmingly positive for a large group of very sick people that the sacrifice does not matter. The end justifies the means in such a case. The opposing religious argument rests on two bases: first, that reverence for human life in the womb is quite explicit in certain biblical references (e.g., Isaiah, Ezekiel). Further, the best-known Christian prayer in popular culture is the Hail Mary (perhaps since it is now used as a general expression to describe an action that is a final effort to achieve victory in either sports or politics), which contains the phrase "blessed is the fruit of they womb, Jesus," so that reference to and reverence for life in the womb is not simply a deduction from theological premises but is built in, as it were, in the biblical and religious account of human life.

The other basis supports the first argument in opposition to ISCR, the slippery slope. Slippery slopes take effect in two ways: one is by habit on the plane of actions, the other is by thought on the intellectual plane. So the objection is both a practical and an intellectual one. Practically, once a psychological barrier is broken, like the four-minute mile or sixty home runs in a season in sports, it becomes easier to go beyond it, and once overcome, the barrier is not likely to be reestablished. A person once having bullied an opponent or cheated on an obligation, that type of action is not only likely to be repeated but to be intensified and become a habit. So also with ISCR—which not only used fetuses for experimentation that were already in existence, as in "snowflake babies" (i.e., unwanted fetuses left over from *in vitro* fertilization techniques for infertile couples), but also created human fetuses *in vitro* for the explicit purposes of using the "fetal material" for experimental purposes—which itself overcame a major barrier: that, is creating human fetuses directly to be experimented on. But the issue lies in where to stop along the trail of normal fetal development. It was pointed out earlier that the argument from human dignity could be ignored as too vague; so also *here*, the assumed barrier is not only vague but seems to lack expression altogether. That is, it is assumed that "of course, we scientists would never experiment on live human babies (in the manner of the Nazi physicians), nor would we ever want to develop a human fetus beyond, for example, the fifth month of pregnancy, when organs, nerves, and so on are developed to nearly human completion to be used for experimental purposes." But then, where is the stopping point; and this unexpressed assumption has not prevented the idea that fetuses aborted voluntarily by women for personal reasons not be used for experimental purposes. The essential difficulty in trying to establish a stopping point on the slippery slope of fetal development is that a strong case for extending experimentation beyond that of a few fetal cells in a petri dish can be made on Utilitarian grounds; what if the sacrifice of a few five month old fetuses can result in the elimination of Alzheimers for millions?

As was stated in Chapter 2, the wall of separation between church and state has always been a porous one that encourages continued attempts at a more perfect secular state by proponents of a secular and scientific worldview. And we see, in the attempts by agents of a secularized state to eliminate any religious argument altogether from the debate over ISCR, an example of pursuit of a "naked public square" (Chapter 4). The slippery slope argument does, however, reveal that some arguments that are legitimately civil, that is, do not refer directly to Christian doctrine or monotheistic religious traditions, nonetheless take a degree of inspiration from religion. Here is where the threads become difficult to pick apart; certainly, human dignity is a concept that makes much more sense in a religious context than a utilitarian one, since it refers to the application of a transcendent aspect to each human individual. Also, the slippery slope argument takes much of its force from the belief that at a certain point, the fetus has become undeniably human, even as it lives within the mother's womb; but it may be said that the hesitation to use identifiably human beings for experimental purposes is a mere matter of sentiment that can, at need, be overcome. To ensure that the resistance to human experimentation is not overcomeable by means of this response, it becomes necessary to have at hand, in these cases, a "bright line" ethical rule—thus far and no farther. (See Chapter 16, on monotheism.)

Notes

1. Coles, *Secular Mind*; 40.
2. James, *Varieties*; 9, 11, 29.
3. Scott Mowbray, "Ignorance: the Cost Goes Up." ("From the Editor" column) *Popular Science* January, 2004; 6.
4. Rustum Roy, 2002, "Religion/Technology, Not Theology/Science As the Defining Dichotomy; *Zygon*; September, 2002; 667–76.

5. For a general account, see "The Stem Cell Debates: Lessons for Science and Politics;" by the Witherspoon Council on Ethics and the Integrity of Science; *The New Atlantis*, #34, Winter 2010; 9–60. http:// www.Thenewatlantis.com/publications/the-stem-cell-debate (accessed 7-1-2012).
6. Bush's speech can be seen at http:// www.Youtube.com/watch?v=EA:Zp5;TO4I.
7. "Bush's limitation on embryonic research were ethical and legitimate—but not, as many observers have noted, anti-science." Brandon Keim. http:// www.Wired.com/wiredscience/2003/03/obamastem (accessed 7-1-2012).
8. Wolfgang Lillje, M.D., http:// www.21stcenturysciencetech.com/articles/winter01 (accessed 7-1-2012).
9. In the highly charged case of children raised by gay or lesbian parents, recent research indicates that such children are not faring as well as those raised in traditional, heterosexual two-parent families. The mere appearance of this research has become a controversial issue. See William Saletan in *Slate*. Ramesh Ponnuru in *National Review*; July 9, 2012; Vol. LXIV, No. 13; 28–30. Original research in *Social Science Research*, by Loren Marks and Mark Regnerus.
10. http:// www. washingtonian.com/articles/health/biotechnology (accessed 7-1-2012).

Chapter 15

Sex and the Secular State

The complete elimination of religiously inspired concepts from policy debate and civil discourse would be problematic, since such bedrock concepts as individual rights and equality before the law owe their origin and force in part to religious sources. It is apparent, nonetheless, that in matters affecting sex, reproduction, and family life, the secular state is fully organized to eliminate any religious connections at all. It is less apparent but no less the case that the secular state ignores scientific reason as well, in the exact sense that the new technologies offered as the result of scientific research are used for ends that are strictly guided by the interests of the state. In these matters, state policies transcend scientific reason to use new technologies of abortion and birth control to inhibit the unwanted growth of the national populations, but it at the same time avoids looking rationally at the consequences of such policies. That is, the secular state is not guided by the scientific research that shows that national populations are in decline, that family life has degenerated, or that the varieties of sexual activity enabled by medical technology have led to increased rates of sexually transmitted diseases. Rather, the secular state pursues two seemingly complementary but actually contradictory purposes—to encourage its citizens to become autonomous selves while increasing its own power to a degree that is detrimental to the well being of its citizens.

Deadly Demographics of the Secular State

This essay has not relied on statistics much, but at this point, some reference to statistics is required to make a salient point. What demographic details point out is the main effect of secular state policy: the decline of populations, especially native populations in many Western nations, and the decline of fertility rates "well behind replacement rate."[1] The United States might seem exempt from this phenomenon, except that most if not all of the increase of our population come from immigration, whether legal or illegal. The population in Germany also undergoes the same effect, with the importation of workers from Turkey and North Africa.[2] The effect is well known and often commented upon, typically by right-wing authors such as Patrick Buchanan, Peter Brimelow, and "Spengler," for whom population decline is the prime indicator of the decline of the West.

The effect, so well established, has several causes we may infer. One cause is that the availability of contraceptives has divorced sex from reproduction, enabling married couples—and unmarried couples, as well—to regulate the number of children they have to one or two or none at all. It is a middle-class phenomenon reinforced by the consumer mentality and the intense kind of individualism currently in effect in Western culture. It is also a statistically observable effect that the more a society becomes enriched and its general living conditions improve from "third world" to "first world" levels, that the fertility rate among (married) couples declines and families become smaller.

Beyond the effective sociological cause lies an ideological cause exemplified in a clutch of state policies found throughout Western nations: encouragement of contraceptive use, validation of homosexual lifestyles, and the legitimization of abortion. The current debate in America at this time surrounds the implementation of Barack Obama's health-insurance law; according to current administration policy, its mandate requires that insurers make available free hormonal contraception, a policy that has raised the ire and motivated the resistance of the bishops of the Catholic Church. (Some critics have claimed that the real object of Obama administration polices is that abortion, and not only birth control, be made available at no cost to women.) However the current debate plays out—upcoming court decisions and elections in which Republican opponents say they will deconstruct the law may stop full implementation—the policies of the Obama administration show what the latest and most advanced thinking is on the part of the fully secularized state regarding sex and reproduction, policies that have direct consequences for the relationship of the individual to society and for family life.

The policies speak starkly about their evident aims, for state support for abortion, birth control, and same-sex marriage all act against the reproductive aspect of sexual intercourse. There is an ecological aspect as well, for there is a prevalent sense among the well educated that the human race is not something special among the species that inhabit the Earth and that are in fact a sort of vermin who take more resources than is justified and turn the flesh of other species into their own food, their hides into leather good, and their native wildness into obedient house pets. From an evolutionary point of view, it is claimed that humans are only the latest small branch on the ever-evolving tree of life, and in the end, there are just too many of them—or us—threatening to overwhelm the planet Earth.

Contemporary speech codes known as "political correctness" have a distinctive and essential role to play in the set of antireproductive state policies, for they enforce the idea that antireproductive policies are good things in themselves. Political correctness is often seen as a laughable crotchet among the overeducated, but its proscriptions have real social force. Opposition to gay marriage is characterized as "homophobia" and pointing out the disadvantage to children of single parenting is characterized as "classism" and an under-appreciation of different "lifestyles." Merely discussing the possibility that there are inherent differences between males and females in mathematics was enough to get the president of Harvard University relieved of his position. Referring to uncomfortable results of antireproductive policies is seen as "extremism," even as the evidence for their bad effects is overwhelming and scientifically based. Who today points out that hitherto obscure sexually transmitted diseases (STDs) have appeared as a result of the prevalence of sexual practices formerly described as "perversions"—AIDS the result of anal sex, chlamydia the result of oral sex. The preferred American solution to these effects is predictably technological, in the form of producing new drugs such as the HIV "cocktail" and a national program of injections to prevent chlamydia infections among young teenagers—boys as well as girls. There is no government program pointing out that certain sexual practices may lead to these diseases; if there were, the response from advocates of non-reproductive sex would probably be that pregnancy is also a disease.

In a sense, the modern state gives its citizens what the ancient Roman empire gave the residents of Rome—bread and circuses. In the current setting, this means welfare entitlements and sexually explicit entertainment, for among other aspects of current state policies regarding sex is a permissive policy toward pornography; it seems that a new form of communication is no sooner discovered and made available by entrepreneurs than it is used for pornographic portrayals. Pornography is now characterized as an "industry," with its participants as "sex workers," and it is constitutionally protected as a form of free speech. What all these policies—easy access to abortion, free contraception, gay marriage, and protected pornography—amount to is a state-sponsored policy that the human race stop reproducing itself.

Such a policy is destined to be self-defeating on several levels, for population decline, if carried out consistently and long enough, will not only reduce the size of the national population but also destroy the national culture that encourages non-reproduction. The self-destruction of the population means the self-destruction of the culture, because eventually there will not be enough of the population, particularly among the better educated, to write, communicate, and think about the culture. Thus the culture will die along with the population, to become something merely of historical interest, something that already appears to be happening as the designated carriers of Western culture, the professors, often take a dismissive attitude toward it as in the denigration of the literary canon as the product of "dead, white males."

Since not everybody shares the state-sponsored policy of antireproductive sex, there will be inherent opposition, which will in effect establish a permanently unsettled minority whose influence will have to be contained. This is the case especially with religious believers, who will, not unexpectedly, feel left out as a community and think they are being discriminated against. But this population of serious religious believers supports the basic activities—not to say *existence*—of the state by their readiness to obey its laws, their willingness to work and maintain themselves and to bear children and form families. Not least among their contributions to the state are their voluntary actions and institutions aimed at the general good, such as hospitals, schools, food pantries, and universities. The contributions of religious believers and of religiously inspired institutions in American society would seem too valuable to disregard or to minimalize; however, the current state policies that encourage antireproductive sexuality place it in direct opposition to the *raison d'etre* of religious institutions. The conflict arises inevitably because laws allowing and encouraging homosexual lifestyles, free access to birth control, and the validation of abortion rights directly contradict the ideal of sexual continence, marriage, and family life that forms a basis of religious morality.

One aspect of sexuality that is implicitly suppressed by the policies of the new secular state is the vast amount of outright sentiment connected to sexual attraction and marriage. Love notes, deep sighs, covert glances, rushes of vast emotion—elation at a potential lover's regard, deflation at their disregard—pet names, declarations of everlasting love, egregious exaggeration of the loved one's positive qualities and a blindness toward their faults; all this sentimental impedimenta of sexual desire and ecstasy leading to hopeful permanence and lifelong happiness is reduced to hormonal chemistry, fatuous words, and reckless descent into a sentimental and sinister pool of unmanaged emotion. But then these sentiments are the ties that bind not only lovers and marriage partners to one another but family units to the society at large. The fragile infant in swaddling clothes whose tender face brings tears to the eyes of her grandparents is the future citizen of her community, producer-consumer of the economy and the guarantor of the health and preservation of the state. A culture of "hooking up" (i.e., of sexual context devoid of sentiment and ideas of attachment) is the inevitable social product of the antireproductive policies of the secular state.

The final consequence of the sexual policies of the contemporary state is that it cannot encourage the building up of personal character as part of its purpose. The manufacturing, so to speak, of persons able to sustain and carry on the natural purposes for which a state is formed can no longer take place. More accurately, the secular state encourages its citizens to always be ready to protect their own individual rights as well as those of others; this alone seems to be the only requirement or ideal that the new secular state requires of its citizens. But, once again, this is a self-defeating policy, since the existence of the state requires an effective number of citizens who will participate in the common life of the state beyond the protection of their rights and provision of their immediate wants, especially on the part of its military, firefighters, and police.

Technology and Guilt

The new secular state, because of its policies that promote non-reproductive sexual activities, takes its stance in immediate opposition to religious tradition; it also relies on a suite of reproductive and medical technologies that are applied in defiance of what scientific studies imply about their consequences. It is not merely that sex has been divorced from reproduction by reliance on birth control, but that birth control itself has been immeasurably improved by medical technology. The lowly condom or "rubber" is still relied upon in many instances (they are typically for sale in vending machines in the men's rooms at bars and night spots), but it too has been improved: Casanova used sheep skin, but modern Casanovas can rely on product improvement based on materials science. The divorce between sex and reproduction has been extended tremendously by the ability to take eggs from fertile women and put them in a glass dish, in a safely guarded medical setting, into which sperm from a male is introduced, the egg actually impaled by a very small needle and the sperm material thus injected. This is a medical feat whose moral consequences are so vast that they are not yet fully realized, for it completely divorces reproduction from the sexual act. This divorce is a social fact that would otherwise be commented on by intellectuals who are normally critical of how technological intervention divorces humanity from nature and pollutes nature itself in pursuit of shallow, materialistic ends; but not much criticism of this sort is to be found.

Because medical technology now provides an effective means of asexual reproduction among human beings, it thereby provides the means by which homosexual couples can have children of their own—or, rather, "step" children, since in many lesbian associations, while the egg is from one of the partners, the sperm is taken from an (anonymous) male. In this way, reproductive technology gives scientific color to the legitimization of lesbian and gay relationships in contemporary society, including marriage. Homosexual couples may no longer be said to be childless as a matter of physical necessity, since they can have and raise children almost the same as heterosexual couples, albeit with required assistance from medical technology. Asexual reproduction also enables heterosexual couples to have children who otherwise could not, providing an alternative other than adoption by which infertile couples can have children who are their natural offspring.

It is noteworthy in the application of this technology that the male is reduced from being marriage partner, husband, lover, and father to sperm donor. Maleness, being a man, has become in our culture less a matter of being someone who fulfills his identity by raising a family or providing for it by earning a living, but rather a feckless figure who has no purpose to his life, not merely subordinate to females, but essentially unnecessary. His cultural identity is locked into a precise range between late adolescence and early adulthood, for in our popular culture, in movies, on television, and in advertising, he is seen drinking beer, making vulgar jokes, commenting on female body parts, and avoiding responsibility—a frat-house existence as depicted in recent movies aimed at young men. One can contrast two current television programs: *Two and a Half Men* shows the antics of irresponsible, unmarried, but grown male figures in pursuit of sexual adventures, and is aimed at a broad audience; *Mad Men* is more of a niche show aimed at culturally sophisticated viewers; it pictures serious men earning a living in a serious industry in the 1950s, as if living in a psychological prison.

Technology permits human beings to avoid the consequences of their sexual actions in a way hitherto unknown in human history. Obviously pregnancy can be avoided or terminated, but another consequence of sexual activity, namely disease, is also controllable in a way hitherto unknown. This medical accomplishment, however, has become a social necessity, because STDs have not been eliminated and the recent popularization of non-reproductive sexual practices formerly only whispered about or available in brothels are now popularized on network television. The response of the secular state has not been to provide warnings of the sort used to discourage cigarette smoking, or programs to discourage such forms of sexual activity among the young, but to support medical research to provide medicines that can relive such diseases.[3]

The final social consequence of state policies regarding sex and reproduction is to alleviate the guilt that often accompanies sexual activity. Sexual guilt is a touchy subject, because while it is seen as a genuinely "bad" thing by cultural elites, such guilt can also be an encouragement to developing a sense of responsibility about sexual activity and its consequences. The former fear of pregnancy has largely been alleviated but not eliminated, while social scientists note what the popular culture reflects, namely the vacuity and barrenness of personal relationships based on passing physical intimacy. On the other hand, the effect of reducing guilt and the sense of responsibility connected with sexual activity has been noted by left-wing cultural critics and seen as a good thing.[4] However, the argument about sexual guilt and responsibility is still being pressed, and it is noteworthy that the one issue that had been seen as to be settled, that is, the permissibility of birth control and the desirability that all children be the result of explicit planning, is now coming to public attention once again. The reappearance of birth control as an issue occurred in 2012 when presidential candidate Senator Rick Santorum mentioned it in his campaign and in the controversy over the Patient Protection and Affordable Care Act (Obamacare) between the Obama administration and the

Catholic bishops of the United States.[5] In effect, these happenings again brought forth the central issue surrounding sex, reproduction, and state policy, namely the divorce of sex from reproduction and the secular state's promotion of non-reproductive sex.

Where Have All the Children Gone?

"Where have all the children gone, long time passing?" the singing group Peter, Paul and Mary once asked plaintively, but it was an anti-war song. Now the question can be asked in the context of in peace, for it is noteworthy not only that there are fewer children than there used to be but that family structure has become far more fragile. This fragility is of major importance to the future prospects of our culture but tends to be passed over when not simply overlooked by intellectuals and commentators although the evidence available from social-science studies is overwhelming. Such evidence comes from the elevated divorce rate, the effects on children of single-mother parenting in terms of increased dropout rates and increased involvement with the criminal justice system, as well as the number of abortions. It is considered unseemly, and possibly "racist," to expatiate on the effect of family breakdown on children, even though it is so fundamental to the well-being of children.

Children are no longer seen as the future of the human race, for any sense of obligation of reproduction is gone from the high culture of America—indeed, it now seems an absurd idea, that an autonomous individual has an obligation to have offspring in order to keep up the population or to keep the culture in existence! Indeed, children have become seen as a burden in advanced thinking; the point is frequently made that, for example, a woman is more likely to survive an early abortion in complete health than by bringing her pregnancy to term and actually producing a child. The cost of raising and educating a child, including college, is often cited and the figures brought forward—in the hundreds of thousands of dollars, approaching a million per child—make it appear that child rearing is an expensive hobby.[6] Children are no longer the seen as the glory and promise of the human race but simply a burden, a choice made by the sentimental and simple-minded or else the result of inattention to proper birth-control methods and accidental loss of "protection." Religious authorities and some politicians have spoken of the "culture of death," but this may refer not just to outright killing of fetal life as to the habit of mind that abhors the making and raising of children, in contrast to all human history and normal human inclination. Here is an account of that attitude by a contemporary French novelist:

> Helene, in truth, didn't have that strong a desire to have a child, and a few years later it was she who proposed that they buy a dog. In a passage where he laments French decadence and a fall in the birthrate (already in the news in the 1930s), the Fascist author Drieu LaRochelle imitates, in order to attack it, the conversation between a decadent French couple of his times, which goes more or less like this: "And then Kiki the dog—that's easily enough to amuse us." She was basically in total agreement with this view . . . (a) dog was as amusing as, and even much more amusing than, a child."[7]

Currently, television advertisements for pet food easily give the impression that pets have replaced children in the affection of adult pet owners, who, in one advertisement, are directly addressed as "pet parents." Madison Avenue understands where the culture is at. But the bill for this attitude is coming due, as government economists continue to predict the failure of entitlements for the elderly, particularly Social Security and Medicare. Before the growth of welfare entitlements for the aged, grown children had responsibility for the care of their elderly parents; they still do in the manner now no longer of taking their failing parents into their homes but by paying income tax and payroll deduction to finance the programs for the elderly; but there are no longer enough grown children to support the programs. All in all, as family life and devotion to children decline, so does the connection between generations; the individual autonomy now so highly prized as the touchstone of delight in our present culture signifies not only our separation one from another but the severing of the connections to prior and subsequent generations. Thus the AARP (American Association of Retired People) lobbies for unsustainable commitments to support the retirement and medical-care programs for the "baby boom" generation without regard to financial burden on their own children.[8]

Explaining the Psychology of State Power

Reflecting on Plato and Freud may seem an arcane conclusion to exploring the topic of the power of the secular state in relation to sex; nonetheless reflecting on the two greatest psychologists in Western intellectual history is a good place to end. For both masters of human psychology, human personality structure is mainly a process of regulating desires—that is, the Freudian id, polymorphously perverse, which Plato, using a mythological term, described as a hydra, a monster with a hundred heads each of which had sharp fangs by which to devour its victim. For both Freud and Plato, controlling internal desires is necessary, because if we do not control our desires, the desires will control us, which leads, in turn, to destructive consequences. It is important to note that in contrast to the usual dualistic analysis of the human psyche—yin/yang, body/mind, male/female, good/evil, –and so on—Freud and Plato put forth a triadic interpretation. Thus the process of control is not a blatant picture of top-down command of unruly appetites by a controlling element of the psyche; rather, for both Freud and Plato, control and regulation of the hydra/id is a complex process involving three elements. And most importantly, Plato and Freud connect the process of internal self-control for the individual psyche to the society in which the individual lives; that is society, in general and particularly the formally organized state.

For Plato, the relationship of the individual to society is by direct resemblance; just as there are three elements in the individual psyche—a rational principle, represented by a human figure; a principle of temper or willfulness, represented by a lion; and the desires, represented by the hydra figure—so also in society. The philosopher king is the executive who relies on the paramilitary forces to regulate the otherwise-destructive urges and political movements of the unenlightened citizenry, which is generally activated by its own immediate self-interest. The picture is an authoritarian and static one, with the necessity of social control paramount since the alternative is the despotic and destructive rule of a majority led by unprincipled demagogues. For Freud, however, the picture is a dynamic one in which the desires of the id are controlled by the superego under the direction of the ego. The ego, however, is the *result* of the conflict between the desires of the id and its repression by the superego, not simply the *manager*. The relationship of the individual psyche to society is direct in Freud's portrayal and not by resemblance, as it is for Plato; for Freud, the superego is not something innate (only the id is innate) but is the psychological impress of the various lessons, impositions, disciplines, and sermons directed at the child by its parents and by society at large, including religious authorities. For Freud, the impact of society on the individual enters the person's very psyche and constitutes an essential part of it. Both Plato and Freud assert the necessity of control over undisciplined desires for the psychological health of the individual and the peace of the state (although Freud is conflicted about it).[9]

Just as self-control is the element necessary for successful living for an individual person, so also regulation of the desires of the state, that is, of organized society, is necessary for the successful life of the state. But what is the essential desire of the state; obviously, it cannot be a matter of physical appetite—food, sex, drink, or fine clothing. Rather it is the innate desire for power, for extension of its control and influence over external events that may affect it, sometimes dangerously, and above all for control over the individual lives of the inhabitants. What food and sex are to the individual id, the very basis of desire, power is to the state; an insatiable need, ever seeking for outlets, ever desiring to expand beyond its natural limits.

Authors such as Hobbes and Adler have designated the need for power as innate to the individual, but in that analysis, Hobbes made that need an excuse for increasing the power of the state to an almost absolute extent, power that must be acceded to by the individual that is limited only when the state desires the death of the individual. Adler substituted power for Freud's sexual desire in his view of personal psychology, but in either case, power, it appears, is the more general and most exact aspect that described the innermost psychology of both the individual person and the state. Power over others gives an individual or a state the ability to take what one wants from others and to alter the social and physical environment in which a person or a state lives. Which in turn means the acquisition of those things, such as food and sex, desired by the hydra/id of the individual person or the wealth of other nations by the state, (e.g., as colonies).

The serious person, or the well-run state, will soon learns the limits of power—that is, the point at which there is a contrary response that forces a withdrawal of claims; understanding the limits of one's own control is a first step to maturity for the individual and security for the state. The infant learns that screaming and holding one's breath may work for a while but after a time is ignored by parents and adults, who become tired of such demonstrations, as the child learns it is better to cooperate with its parents rather than to try continually to impose its will on them. Augustus, the first head of the Roman empire, learned through bitter losses of its legions on its borders that the empire could not extend its power, for example, into northern Europe, and after a while, possession of other lands becomes more costly than it is worth (as the Soviet Union and European colonial nations also discovered).

Constant attempts to expand one's power on an individual and state level also has the consequence of unsettling the person or the social entity within themselves; the constant scheming to get attention and have one's way uses up a lot of psychological energy on the part of an individual and roils the internal politics of the state. A person always measuring herself against others' beauty, success, or power is not likely to have a happy life and will almost certainly deny herself inner peace. The cost of competition for control of the Roman state was civil war among its leading generals, Caesar against Pompey, Octavius against Marc Antony.

But the necessity for a mechanism of control of the appetitional desire for sex and power is a psychological fact often denied in American culture these days. Rather, self-expression and a full flowering of one's inner being, as if nothing felt should be inhibited from explicit expression, is the rule and felt to be the key to ultimate personal happiness. One of the benchmarks on the way to the full development of this aspect of current American culture was the appearance of a woman's health and lifestyle magazine called *SELF*.[10] It is still in publication, and it is the forerunner of many other such magazines (and now websites) expressing the same point of view. And now there are men's magazines and sites expressing the same point of view—that is, that the principle aim of healthy and imaginative living starts with the personal self; its mantra is "if you do not love yourself, you cannot love others." By contrast, there are no popular sites or magazines called *Other People* or *God*. It seems that the only figures in our popular culture who speak about any moral imperative to devote our energies and attention to others are professional athletes and their coaches; "no 'I' in team," as the saying goes.

Nevertheless, inner controls there must ultimately be in order for a person or a state to live at peace within itself; Plato and Feud elaborate a three-part description of the human psyche in which the rational ego is hopefully in control of the hydra/id as it applies the psychological mechanism necessary to keep it in check, resulting in a balanced personality. But what, if anything, can provide an inner check on the hydra/id of the state, of the desire for power that, to paraphrase Hobbes, ends only when the state comes to its end, and like the Roman empire, dissolves into its constituents? Hobbes applied his analysis of power to individual persons but also thought that the power of the state should supplant the otherwise destructive desire for power among its citizenry. What Hobbes saw and experienced was a savage civil war in England only recently concluded, and his solution for preventing civil war was that the state should have a monopoly of power. The state's monopoly of military power was necessary for peace, but a state monopoly on religion was necessary as well, for it seemed to Hobbes the force most likely to destroy the state from within. Now, in the early twenty-first century, our situation is quite different in that state, while not controlling religion has largely succeeded in establishing the "naked public square." Only now we are left with serious concern not about the desire for power of individuals or of the church, but of the state itself. The hydra/id of the secular state is not controlled these days either by science that predicts the dire demographic consequences of its policies or by the bright lines of ethical morality available from religious teaching. In the next chapter, the latter issue is taken up.

Notes

1. See Russell Shorto, "No Babies?"; *New York Times Magazine*, June 29, 2008. http://www.nytimes. com/2008/06/09/magazine/29birth-t.i (accessed 6-1-2012).
2. Giampola Lanzieri, "Population in Europe 2007:first results"; http://ppeurostat.eceuropa.eu/cache/ITV_OFFPUB/KS-SF-08 (accessed 6-1-2012).
3. But see the CDC report on the greater likelihood of STDs arising from oral vs. vaginal sex. www.cdc.gov/std/stats11/toc.htm. Reported in *Salon*, (August 18, 2012).
4. See Michael Lind, "What Killed Social Conservatism" *Salon* on-line magazine, 5-28-2012. www.salon.com/2012/05/22/what_killed_social_conservatism.
5. US Conference of Catholic Bishops, April 12, 2012; "A Statement on Religious Liberty." http://www. USCCB.org/issues-and-action/religious-liberty (accessed 7-2-2012).
6. There are a number of websites that enable the calculation of annual costs. See the US government website, www.cnpp.usda.gov/calculatorintro.htm.
7. Michel Houellebecq, *The Map and the Territory* (trans. Gavin Bowd); New York, Knopf, 2012; 188.
8. AARP, "Myths versus facts"; http://convio.net/site/PageNavigator/Myths_vs_Facts.
9. Plato, *The Republic*; Book IV Sigmund Freud, *Civilization and Its Discontents* (trans. J. Strachey); New York, Norton, 1961; sections VII, VIII; 70–92.
10. See the *SELF* website; www.self.com.

Chapter 16

Inevitable Monotheism: Why God Hasn't Altogether Left the Public Square

Even as advocates of complete separation of church from state sue local towns to eliminate crosses from war memorials and remove stone tablets with the Ten Commandments carved on them from public lands, evidence of the presence of religion in the public square persists. Political candidates for the highest offices in America must, it seems, loudly announce their belief in God along with their devotion to mom, the American flag, and apple pie. On the right, Republican candidates court the evangelical vote, while on the left, while reliance on religious support is eschewed, Democratic candidates do occasionally attempt to align their policies with explicitly religious doctrines. As the removal of explicitly Judeo-Christian doctrine from public life proceeds, it becomes apparent that certain ideals that have been taken for granted in fact are based on a religious sensibility, including devotion to the family, reverence for human life, simple courtesy to others, and sexual modesty among them. In short, the public square is not completely naked, not yet anyway, and it is becoming increasingly apparent that complete elimination of a religious sensibility based on traditional Christian and Jewish and, perhaps in the future, Muslim beliefs and practices is not really feasible or desirable. This chapter attempts to explain why, as a cultural phenomenon, this is so and, frankly, at the same to time to try to demonstrate what benefits belief in the One God provides for an explicitly secular society. Of the three fragments of religious belief described in Chapter 12, monotheism appears to be the most resilient and useful in the long run.

Psychology and Reality in Religious Belief

The general secular explanation of religious belief is that it is a form of psychological pathology; William James begins *The Varieties of*

Religious Experience with this explicit understanding.[1] Accounts of the lives of the mystics, who are those religious believers who manifest religious belief in its purest form, might seem to prove the connection between religious experience and psychopathology. There is an entire set of female mystics who hear voices, see visions, perform fasts and penances, are celibate, and who provide expressions of simple piety, who therefore seem to provide examples of religious faith as a hysterical phenomenon—Gemma Galgani, Catherine of Genoa, Catherine of Sienna, Joan of Arc, Mary Baker Eddy, Mother Annie Lee, Therese of Liseaux, Theresa of Avila, and Mother Teresa. Nor is this limited to figures of historical note, as common experience shows examples of elderly women at their beads or constantly reading the scriptures, working at volunteer positions in their churches and temples, praying more often than is seemly, and using religion as a source of personal strength. The attraction of religious belief for women has often been noted, which serves subliminally to place religious belief in the context of a psychological refuge for the weak of mind and will.

This explanation results in an ugly stereotype, which has been decried by feminist scholars who have attacked the identification of women's religious belief with hysteria as a form of gender prejudice.[2] In fact, they could reply, there are many examples of male mystics, indeed a long line of them, including Paul of Tarsus, Bernard of Clairveaux, George Fox, Ignatius of Loyola, Muhammad, Joseph Smith, Pope John Paul II, and again, many others of less note who can be commonly observed serving as deacons of their local churches, reading the scriptures, making a minyan, and so on. Considering the mystics as examples of religious faith in its purest form seems to imply that religious belief in general is obviously akin to psychopathology, but here we come across a contradiction. The functional description of psychopathology is that it results in the patient's inability to cope with reality; in the case of religious mystics, this means relieving inner stress with inventions of invisible voices, visions, and supernatural manifestations. Patients who manifest such pathological and hysterical tendencies are typically unable to function in daily life without a large amount of counseling, drug therapy, and personal support.

Yet, the toll of mystics above contradicts this portrayal, for they were not only religious mystics but were forceful and effective figures. Theresa of Avila was a founder of convents and a successful reformer, Joan of Arc was a successful military and political leader, Catherine of

Sienna convinced the pope who was residing in Avignon to return to Rome, and Annie Lee and Mary Baker Eddy were founders of religious denominations. The practical effectiveness of the male mystics, including St. Paul and Muhammad, do not need to be recounted, because they are world-historical figures. In all these cases, the mystics not only have intense religious experiences but also provide charismatic leadership and engage actively in daily life, constantly making decisions while bringing to successful conclusion long-term strategic visions. There can be no serious allegation that their religiosity is a form of mental pathology or that their personal psychological structures are abnormal in these cases—except for their exceptional strength.

These examples of mystical religious faith indicate that religious faith in general is not an indication of psychological weakness but on the contrary is often the result of exceptional strength of mind and will. Nonetheless, it is broadly accepted that religious belief is a form of irrationalism and mental cowardice, as stated recently by Hitchens.[3] It is only the strong minded, modern writers have said, who can endure the facts that existence does not have any meaning, that the universe gives no indication that it has a purpose, and who recognize that there is no God. It is only the strong minded and intelligent who can deal with the fact that there are no absolutes in life, and it is the weak minded who seek refuge in the notion that there is an afterlife and a loving God who will dispense reward and justice. But this description is a canard, even though it has become widely accepted in modern thought and which not incidentally is self-complimentary to those who criticize religious belief in this manner.

This portrayal is false when measured against many people's personal experience—of the faith-filled grandmothers, for example, who keep two sets of dishes or attend daily mass or are constantly reading the scriptures, who are the backbones of their extended families, even as their feckless grandchildren become assimilated in American consumer society. It is often religious figures, even someone who excited such disdain as the Reverend Jerry Falwell, who appear to be a persons of strong will and resolve in the face of constant opposition, or the diminutive figure of Mother Teresa who spoke against abortion in the presence of President Clinton and in favor of compassion for the suffering to President Reagan; both Falwell and Mother Teresa were founders of successful religious communities. On the other hand, it seems it is often those who are less organized in their psyches who leave their family's religious faith and who experience long-term difficulty in their personal lives, such as divorce and going from one thing to another; at least it has been so in this writer's observation. On the epistemological level, it seems as if in accepting doctrines that are miraculous, such as Smith receiving the golden plates, the resurrection of Christ, Moses parting the Red Sea, or the flight of Muhammad to Jerusalem, that it is the stronger personalities who can accept these doctrines, and the less psychologically strong who are discouraged by them. In which case, there is a reversal of the stereotype that religious belief indicates psychological weakness and that lack of it indicates independence and self-assurance.[4]

The engagement of a person acting in a religious-belief system includes a degree of intensity that seems to imply that the believer is making his or her own reality through force of will. However, the observable fact is that engagement in self-defined realities is a strategy that ultimately leads to disaster: the arrogant general who fails to credit the ferocity of his enemy; the scientist committed to a research program who refuses to recognize contradictory evidence; the businessman convinced of his sales strategy, who is unable to cope with changes in the market. And surely, the example of people whose strength of conviction gets them into unnecessary problems and conflicts is known to us all; we may even be them. We can conclude that while there is a psychological element of willful creation when anyone is convinced of a truth and proceeds to take action within the commitment to a belief system, that eventually the element of reality or lack of it will have an effect, positive in one case, negative in the other. The angry person accusing his friends of deceit will either alienate his true friends or expose a malicious conspiracy, and while there may be only a partial truth to the matter, it will in the end be mostly one thing or another.

As these reflections affect religious belief, we may imagine that the interaction of belief and reality will have a unique effect. In religious belief, the "success" of engagement in the belief system at first is usually internal, resulting in an increased sense of assurance, resolve and ability to overcome difficulties and temptations. Internal success then encourages religious belief to express itself in external acts of zeal and charity; such efforts may or may not be well received, but they will have an effect. Over time, the religious motivation will be tested, and if sincere in her religious belief, the believer will eventually discover an effective and personally satisfying means of expressing her belief. But such a reinforcing effect is ultimately dependent on the element of reality in the venture of religious belief. But if religious belief does not create its own object of belief, then its object, which in the West we call "God" (or its equivalents, "Dieu," "Gott," etc.), may be presumed to be responding to the venture of belief on the part of the religious believer; or, in other words, if you love God, he will love you back. Religious belief is not a success philosophy, for often the dictates of divine love will often ask incredible endurance from the believer, of a sustained lifelong effort in very difficult circumstances, efforts that in the worldly sense may bring not success but failure. And yet such efforts, as in the case of martyrs and long-suffering spouses, are often successful in their own, if not worldly, terms.

The Dangers of Monotheistic Displacement

There is, apparent throughout all human history and in most human psyches, an unrelenting need for the knowledge and experience of the transcendent. References are plentifully scattered throughout human history of mankind's need for the knowledge of God and is a universal theme of all religious writing. This need for an experience of divinity was poignantly expressed in the prayer of St. Augustine, "Our hearts are restless Lord, until they rest in thee." It may be objected that many intellectuals in the West these days do not feel a need for an oceanic sense of religious completion and indeed are busy rendering critical attacks on the need for religious completion, which is described variously as a psychological weakness, an irrational impulse, or a threat to civilized life. These critics point out that the desire for religious completion is, in certain cases, not limited to reforming the individual religious believer's own self but extends outward to the social realm. Certain religious believers not merely share their faith, however, or limit themselves to performing charitable works, but attempt to coerce compliance to religious norms by means of political power. Thus critics are warning of the dangers of religious belief, which are apparent in right-wing domestic politics and Islamic fanaticism in foreign affairs.

But there is a larger and more opposing danger, namely that of *monotheistic displacement.* If we follow the teleological account of human nature from Aristotle to some contemporary psychologists, it appears that nothing in human nature is there but for a purpose or to fulfill a function. Hunger manifests a need for food, without which the organism will die, and thirst manifests the essential need for water. In extreme cases, organisms will take in substitutes when denied such essentials, eating sand if food is not available and drinking salt water when dying of thirst; in the case of sexual desire, humans may sublimate with such activities as sports. But sand will not do for food and toxic liquids for water, and the organism taking them in will surely die; in the social context, denial of sexual contact by too many people or interruption of the coital act by birth control will result in the decline of a whole population. So also, to complete this step of the argument, if the religious need for a sense of wholeness is not properly directed to the knowledge of the One God, then substitutes for religious belief will be substituted which will result in harm to the individual human being or society as a whole. There are outstanding examples of the maligning effects of monotheistic displacement in recent times.

The need for religious wholeness in most people is only rarely displaced by replacement myths or substitute systems of ethics but by a concentration on fulfilling appetites. Of these people, St. Paul said that their god is their belly. Their god may also be placed a little lower on their anatomy, for erotic excitement can engage attention and provide distraction as well as provide a temporary sense of fulfillment that is really a form of exhaustion. However, sexual desire can fill the mind and heart, and history provides many literary examples of the exaltation of romantic/erotic love from the Song of Solomon to the novels of D. H. Lawrence. Indeed, St. Paul uses the image of marriage to provide the model for the relationship of Christ with the church, calling it "the bride of Christ." Such exaltation of erotic desire is nonetheless a form of monotheistic displacement if the imagination and intellect stays on the erotic level, for not only can promiscuity become boring, but ultimately spouses die and love affairs break up. Thus the fulfillment of the appetites, for food and sex, but also the appetites for success, beauty, and glory provide only a false and incomplete substitute for the sense of wholeness provided by belief in God and engagement in the realm of the spirit. Reliance on substitutes for belief in God leads to bullying others, depression, and a frenetic doubling of useless activity. Such displacement is not worthy of the dignity of the individual person and is necessarily harmful to the state, because it increases discord among its citizens.

Twentieth-century observers confronted with the awful and overwhelming fact of totalitarianism often made the observation that the ideologies of communism and Nazism were religion substitutes.[5] This was apparent in the May Day demonstrations held in Red Square during the Soviet era, with massed crowds watching rockets, tanks, marching soldiers, and red flags giving testimony to the power of the state and the victory of materialism over the consolations of the spiritual life. A special aspect of these May Day demonstrations was the huge flags that were carried, on which were portrayed in heroic manner the prophets of the new era: Marx, Lenin, and, for a time, Stalin. But surely the display of these huge portraits was based on memories of the religious icons of the Virgin and her Son carried in the religious processions of the Russian Orthodox Church, so that in this way the displacement of religious belief into the political realm of Soviet Communism was made explicit for all to see. That the body of Lenin is preserved in a mausoleum reflects plainly the miraculous examples of the bodies of holy people that resist corruption that are part of Christian tradition. Even the hats of the Soviet military officers, which were large by conventional standards of military dress, were reflections of the miters and head dress of Russian Orthodox priests and bishops.

The Soviet ideology was explicitly atheistic so that the displacement of the human need for wholeness from God as its proper object to the political regime was indirect. However, the fascist and Nazi regimes were not atheist but deliberately attempted to replace the Jewish and Christian religions of their populations with makeshift constructions of their own. Thus the reversions to pagan pre-Christian symbols not only brought forth for Mussolini the memory of ancient Roman grandeur but was an attempt to displace reverence for the Church with reverence for the fascist state. The resurrection of the ancient Roman symbol of the fasces, the bundle of sticks tied into a rough cylinder in the middle of which was placed an axe, was meant to displace the Christian cross. The Nazi approach was cruder, as it reverted to pre-Christian German tribal religion and so took the swastika from the worship of the ancient Aryans to replace the Jewish torah scrolls and the Christian cross. In all totalitarian states, even as they attempt to destroy all effects of monotheistic worship, devotion to the leader forms a kind of religious cult that provides a unitary focus for popular emotion (as the current example of North Korea indicates).

The displacement of the religious impulse to search for wholeness and completion to the political realm was an undeniable element of totalitarian psychology and perhaps at the root of the totalitarian phenomenon. It can be speculated that the immersion of the population of Europe in 1900 years of Christianity had given European culture the sense that human existence could be fashioned in a way to transcend the tedium of day-to-day living, and that human life was meant to be infused with a sense of divine purpose and grandeur. But once identification with the Christian faith became only nominal for large parts of the European population during the nineteenth century, the appetite for the sense of completion and purpose remained frustrated until filled by ideologies that magnified the state, the party, or the race above all else. As the history of the twentieth century indicates, the effects of totalitarian domination were brutal in the extreme and amounted in practice to a derogation of human life and dignity such that whole populations were treated like cattle. Nations giving in to what Arendt called the "totalitarian temptation" is an experience that—God willing—mankind will never have to endure again.

Monotheistic displacement takes place not only on the political level but on the intellectual level. Philosophical and scientific critics of religion not only find contradictions and purported fallacies in religious belief but very often offer what are, in effect, religion substitutes. In the early twentieth century, Bertrand Russell and Alfred North Whitehead composed the *Principia Mathematica*, which for the first time, combined the basic laws of mathematics with those of logic in formal, consistent terms. Their monumental effort proposed a vision of comprehensive meaning based on formal structures of thought, a field that had undergone tremendous development in the late nineteenth and early twentieth centuries. Their effort is an example of how intellectual critics of religion feel compelled to offer, if not a direct substitute for religion, something that nonetheless fulfills the need to apprehend the universe and all experience with a sense of the whole. Not incidentally, the fate of Russell and Whitehead's great formal system was darkened by the discovery (by the mathematician Kurt Godel) that high-order, complex systems of logic such as that of the *Principia Mathematica* could not be proven self-consistent and thus must remain logically incomplete.[6] Russell went on to other non-logical pursuits, including writing a popular history of philosophy and, more relevantly, advocacy of various social causes including free love, open education, pacifism, and nuclear disarmament; Whitehead went on to become the last of the great metaphysical system builders of modern philosophy. For both men, it seems as if the complex logico-mathematical system they created was a displacement of the sense of wholeness usually sought in religious belief.

In the American cultural context, the removal of the worship of the One God from public affairs may not result in a more perfect secular union, but contrarily may cause the formation of a new religion in America, namely worship of America itself. The religious aspects of "Americanism" are already in place in the reverence for the office of the presidency and the Supreme Court (if not the Congress), for the American flag, in the presence of state monuments and war memorials, in the veneration of state texts (not necessarily the Constitution these days, but definitely including the Declaration of Independence), and not least of all in the belief that all Americans seem to accept that America is a special nation. The sense of American exceptionalism does not exist only on the Right but also on the Left. On the Right, and in the general population, Americans tend to believe that our nation is a city on a hill, designated by God for a special purpose. On the Left, the secular sensibility sees American history in terms of a progression of increasing freedom for oppressed minorities including slaves, women, the laboring class, and now homosexuals. The idea that Americanism is a religion in itself has been made explicit by David Gelernter, a well-known essayist, who describes it as the fourth great religion (Protestantism, Catholicism, Judaism, and now "Americanism"). He traces the religious elements in America's history from the Puritans to George W. Bush to show not only America's religious origins but to promote that idea that America itself shares the essential elements with the other three great monotheistic religions, including missionizing and belief in the One God.[7] But if God is removed from Gelernter's equation what remains is America alone, as a divine entity.

The Advantages of Bright-Line Ethics

As a thought experiment, consider that you are driving on a lonely road late at night in the middle of a rainstorm and that your car breaks down. There are only two numbers in your cell phone: one the number of a renowned atheist, Sam Harris or Richard Dawkins, the other of a famous Christian conservative preacher, Billy Graham or the late Jerry Falwell. Then ask, whose number will you press? That is, who is most likely to answer your need for help? The answer is the Christian conservative, who, whatever objections one may have to his preaching (paternalistic, homophobic, etc.), is more likely to send out help or come himself in answer. This is not to argue that Dawkins would not help you and that Falwell would have, but that in general you are better off calling the self-professed Christian rather than the self-professed secular atheist. If the Christian refuses to help, you can refer to Jesus's parable of the Good Samaritan and his command to do unto others as you would have them do unto you. To refuse you assistance at that point would mark the Christian as a hypocrite. On the other hand, there are few responses you can make to the secular atheist; what could you say in response to his refusal to help to the man who wrote a book called "The Selfish Gene"?

It is a claim often made against religion that its laws regulating human behavior are restrictive and harsh, destructive of human freedom on the personal level and of social peace on the community level. From the anti-nomian point of view, which is dominant in today's culture, religious believers do not have the right to make laws or try to influence the political process in any way based on religious commandments. From the anti-nomian point of view, there is only a difference in degree between a group of local ministers agitating for low-income housing and the Taliban enacting the death penalty against homosexuals. In all cases of attempts by religious people and organizations to influence civil law, the anti-nomian impulse is to see only the potentially destructive effects of the most extreme examples of prescriptive moral laws characteristic of religious ethics. Thus it is common that any organized opposition by religious believers to egregiously violent or sexual entertainment is always characterized as an attempt to circumvent the separation of church and state and to suppress the first-amendment right of freedom of expression.

Religious ethics do have a quality of rigidity about them. They are rules presumably given by God to the human race and thus brook of no objection; further they do not include the "wiggle room" human beings usually seek in prescriptive moral laws. But this rigidity is an advantage of religious ethics, since it provides a "bright line" by which to evaluate human actions on the ethical plane. Religious ethics, because they are absolute, also tend to provide a clearer and higher ideal of human behavior than those that natural philosophy is likely to decide on the basis of human intelligence alone, that is, unaided by revelation. By excluding divorce, euthanasia, or abortion, or by attacking the capitalist system as harmful to the poor and socialist regimes for suppressing human freedom, religious ethics renders a portrait of human life as it ought to be lived in the ideal. We may react that such an ideal is unrealistic, but by being aware of the ideal standard, we are more likely to live less ethically slovenly lives, both personally and communally.

To take a most difficult case, what to do about an ancephalic infant, of whether to provide the minimum of food and water to the little being or simply euthanize it—or let it expire on its own. In the ethical philosophies of Aristotle and Peter Singer, the obvious policy is to let the little being expire on its own and give it no water or food. It appears to be an extreme position to argue that the newborn should be given the minimum assistance for life, and it seems a given that only a religiously inspired ethics could prescribe such a policy. Nonetheless, the persistence among religious authorities to make the case for minimum life assistance encourages respect for the deformed infant and encourages in a larger degree respect for human life in general. Religious ethics demands that we recognize, beyond the merely poetic or sentimental level, that the ancephalic infant is still a member of the human race, deserving of what humans at a minimum should be provided.

The necessity of a religiously revealed bright-line ethics is apparent in a time when biological theory and practice tend to reduce the human essence to its evolutionary origins, deleting where possible any reference to distinctively human traits beyond the animal or mechanical. Evolutionary theory has been applied to explain the existence of the ethical systems that are common to human cultures as a selective strategy for transmitting genes from one generation to the next. Mother love, patriotism, the heroism of first responders, common courtesy, Christian charity, Jewish zeal, Buddhist compassion, and Muslim dedication to the poor are all reduced to a hidden means for passing the individual's genes onto his or her offspring.[8] Genetic technology reinforces this theoretical tendency by demonstrating in how many ways the human genome can be manipulated, making it seem as if the human essence is entirely a material thing that can be reduced to laboratory procedures. The reduction of the human essence to its biological origins makes it possible to argue for euthanasia and infanticide, and it is in this context that religious ethics is necessary in order to elevate human nature above animal nature and to protect fetal life from laboratory manipulation. Religious ethics recognizes that human beings are not only biologically more complex than animals but have moral responsibilities that surpass the animal struggle for existence.

It is not only in its prescriptive moral laws, however, that religion makes the case for an elevated view of human nature, for its portrayals of saints, martyrs, confessors, and missionaries, as well as other holy people, provide examples of how life on the ideal plane can be lived. Granted, we may be put off by the zeal and uncompromising charity of Albert Schweitzer or Mother Teresa, who gave up their lives for a cause, but we can also be ennobled by their examples. Examples of such zeal and charity do not arrive as often or with such intensity from any source other than religion, that is, from believers actuated by their acceptance of revealed religious ethics. True, people are frequently actuated to noble and self-sacrificial acts by other sources such as love of family, devotion to a social cause, patriotism, or self-regard (who wants to be seen as a coward if it can be at all avoided?). But by a large margin, the most constantly noteworthy examples of exemplary self-sacrifice come from religion.

Without an external, authoritative source of ethical doctrine, we are thrown on our own moral and intellectual resources, and it is an inevitable part of human nature that we will fashion standards of ethical behavior based on how we actually act or think ourselves reasonably capable of acting. We are unlikely to specify a high standard for ourselves if we do not as a matter of course keep our marriage vows, think of others first, drink alcoholic beverages moderately, or keep our tempers when provoked. There is always and in every circumstance a human tendency toward self-justification, and humanly created ethical standards will reflect this; otherwise, we put ourselves in the place of living with perpetual, self-made, and unresolvable guilt.

While it is often said that human beings can be good without the support or guidance of religious rules, and we often do indeed meet such people, the fact is that they are usually of an even temperament, socially well adjusted and secure in their own environment. Some people are by nature quiet or soft spoken, not contentious or otherwise inclined to bother other people, but preferring to be cooperative. It is not therefore of great moral credit to them that they do not commit unethical acts as a rule, since they do not have to work hard to fit their behavior to the norms of society and be known as a "nice" person. On the other hand, we look with favor on people who suffer from personality deficits such as depression or alcoholism, or who have been ignored or abused as children, but who have overcome these personal obstacles to become functioning people and find their places in society. And very often people who suffer such personal affliction rely on their religious belief for inspiration and support. Further, the ethical standards that come from nonreligious sources usually do not extend beyond what is perceived to be obvious or necessary, while religiously inspired ideals can reach beyond the quotidian measures of what constitutes ethical behavior to radical or heroic levels. The agitation against slavery in American and England in the nineteenth century came largely from religiously inspired people, and the civil-rights movement of the twentieth century was led by a succession of inspired Protestant ministers including the Reverend Martin Luther King Jr.

In Chapter 4, responses to the decline of the American religion were considered. Neuhaus asserted that the overall result of the decline was a degeneration of our public life, while Wolfe saw the decline in positive terms. But Wolfe's response can only apply on an ideological level; because he assumes that fundamentalist religious belief is bad for democracy, he thinks therefore that the decline in religious belief is a good thing. As was pointed out in that chapter, Wolfe does not consider the practical rather than the ideological effects of religion's absence from the public square. The social process by which secular standards of behavior have replaced those promoted by traditional Protestant and Catholic versions of the Christian religion in America have resulted in a measurable decline in what may be loosely termed "public morality." Crime rates have been going down lately, by means of comparison of the crime rates of the first decade of the twenty-first century with the 1970s. However, when the comparison is extended to the 1950s, the rise in crime and the associated decline in public morals are undeniable.[9] In fact, the difference is so large that it signifies a toxic sea change throughout our society with regard to ethical behavior that it constitutes, it may be fairly said, a crisis.

Beyond the area of criminal law, it is often noted today that common courtesy and self-restraint are no longer valued in our culture, since regard for others has been replaced by elevation of the autonomous self as the aim of public policy, the glorification of "idols" and athletes' individual accomplishments in popular culture, and the unmitigated desire by Wall Street's "masters of the universe" to accumulate as much wealth and power as possible. By contrast, religious ethics promotes humility, compassion, self-sacrifice, and service to others, not as a matter of recognition of individual rights but as a voluntary giving of self. Religious ethics are not based on legal requirements or a theoretical contract made under such abstract constraints as Rawls's "veil of ignorance." It is becoming increasingly apparent that contractual civil ethics is not enough to ensure a level of common ethical behavior that makes life livable in a modern democracy.

Notes

1. James, *Varieties*, on pathology; 8, 9.
2. Rudolph M. Bell and Mazzoni, Cristina, *The Voices of Gemma Galgani*; Chicago, U. of Chicago Press, 2003; 227–231.
3. Christopher Hitchens, *God Is Not Great*; New York, Twelve, 2007; 7–13.
4. Eric Voegelin, *The New Science of Politics*; Chicago, U. of Chicago Press,1952; 122.
5. See Voegelin, ibid.;132, 179. Hannah Arendt, *Totalitarianism*: New York, Harcourt Brace Jovanovich, 1968; 43, 44, 160, 161.
6. Godel's (very technical) paper is reprinted in Jean van Heijenoort, *From Frege to Godel*; Cambridge, Harvard U. Press, 1967; Kurt Godel, "The completeness of the axioms of the functional calculus of logic"; 596–616.
7. David Gelernter, *Americanism: The Fourth Great Western Religion*; New York, Doubleday, 2009.
8. There has been a flurry of sociobiological explanations of religious belief recently. See Pascal Boyer, *Religion Explained: The Evolutionary Origins of Religious Thought*; New York, Basic, 2002. But see an early attempt in Edward O. Wilson, *On Human Nature*; Cambridge, Harvard U. Press, 1978; 169–194.
9. For comparative crime rates, see http:// www. jrsa.org/programs/Historical. pdf (accessed 7-3-2012).

Chapter 17

Cultural Disunity, Toleration, and the Secular State

The argument of this book has become complicated, a fact that reflects the complex nature of the culture of America at this time. While the treatment has dealt with the formative influence of religion and science, this is not meant to deny that other influences are at work in the culture or help to define it, including popular entertainment, late capitalist economics, and an attenuated revolutionary politics. Furthermore, while religious belief has declined and the Christian faith no longer provides a fundamental basis of American democracy as it once did, science has not replaced it. The belief that science is the engine of progress in the Enlightenment sense cannot provide the justification for the secularization of society, because modern science is no longer presumed to provide absolute truth. The beneficiary of this double breakdown is the new secular state, for without either religion or science as the intellectual justification of its policies, the empowerment of the state itself has become its *raison d'etre*. The book's conclusion is a good place for some final thoughts on the issue of cultural unity and disunity, a theme that has provided this book's background. The disunity of American culture was fully displayed in Newtown when President Obama, whose administration has been termed by a Catholic writer "the most aggressively secular . . . in American history," relied on expressions from the Christian gospels to express his sympathy and to provide consolation to the residents and to the nation.

Stages of Secularity

By now it has become necessary to try to define secularism more exactly, as the subject has not only become popular but has complexified, appearing in a variety of social and historical circumstances and so penetrating the culture. A useful approach, by the scholar Abdel

Wahab Elmessiri, is to distinguish "partial" from "complete" secularization, which suggests a plausible distinction between formal government policy on one hand and what takes place in a more general social environment on the other. That is, the separation of church and state (which also guarantees freedom of religious worship) in the First Amendment of the American Constitution is an example of partial secularization. A clearer example comes from England during the nineteenth century, at the start of which there were legal barriers to full participation in civil life based on a person's religious identity, which prohibited voting rights and entrance to Oxford University, for example. Since there is an established Church in England of which the monarch is the nominal head, action by the British Parliament was required to lift the barriers against, in succession, Protestants who were not members of the Church of England, Jews, nonbelievers, and Catholics. Such example of partial secularization are, in effect, the lifting of legal barriers to allow the inclusion of people of different religious beliefs to more fully participate in the civil life of a nation.

Secularization in the context of the twenty-first century seems to mean something else, however: a kind of *complete* secularization in which religious doctrine and practice is removed from the informal but omnipresent forms of social thought; the mores, in fact, of a population. It is a great mystery, really, how this has come about in the West and in America particularly, and as a result, many articles and books have been written about, and much discussion and thought devoted to, the phenomenon of complete secularization. It is generally assumed that the cause of decline in religious belief and practice in the West is due to causes external to religion itself, yet this is not entirely so. On the contrary, it appears as if one cause comes from the decline of religious fervor among religious believers themselves; whether this, in turn, has been caused by the pressures of social disintegration and rapid changing of social structures is another question. However, just because religious faith is opposed in its essence to "the world, the flesh, and the devil," practicing one's faith also means applying it outwardly beyond personal prayer and worship within the walls of a house of worship to the ambient social context. Thus understanding a radically changing society affected by new technologies and extended forms of bureaucracy, and then applying religious norms to the emerging social context (which is an ongoing process, in any case) has been made much more difficult given the fundamental changes in Western culture over the last century or so—that is, since the time of the First World War.

A usually unmentioned cause of secularization is technological improvement. Communication technologies, to take one example, have transformed normal life in profound ways; it is not just that Twitter or iPads have recently emerged; it is that more fundamentally since the time of the telegraph, instantaneous transmission of language over great distances has become possible and now commonplace. Without the mediation of time and distance, communication has not only become more rapid but seemingly changed in nature, requiring less thought and preparation. And in this time of late capitalism, it seems as if the main function of new communication technologies is to provide a means of advertising consumer products, for advertising has become an omnipresence which has taken over radio, telephones, television, and cable broadcasts. Christian advocates often summon up the example of St. Paul, the great communicator of the Christian message of the first century, imitating him by using modern means of communication to advertise the Christian message in our own day. But perhaps as in St. Paul's day, it is very difficult to get one's message across given the competing and discordant messages that overwhelm the media environment. In any case, the Christian message and religious understanding in general are no longer the sole or the main voice one hears on matters of faith, morals, or public policy; indeed one has to search in the mélange of messages, data, and claims for attention to find a legitimate expression of the religious point of view. Consequently, religious believers become discouraged, no longer seeing their faith reflected in the media environment, making their faith seem irrelevant, thus making it necessary for religious communities to close in on themselves in order to secure a sense of common belonging and to retain doctrinal purity. In this instance, secularization has come about as an unintended consequence of the communications revolution.

There is discernible a third stage of development of secularization, however, one that proceeds beyond *complete* secularization, for that stage can be thought to take place purely on an informal level, as a result of various social and intellectual movements such as the advance of scientific ideas or of consumer culture or political trends toward individualism. A third level comes when the governments of Europe and North America reenter the process of secularization, a stage that can be termed *state secularization.* As can be observed in the current political situation in Europe and America, state policies are no longer neutral concerning religious belief, leaving it, as states formerly did, as the prerogative of individual citizens or of centuries-old religious institutions whose historical origins predate the formation of existing states; rather the influence, doctrines, and practices of religion in general are now directly opposed and moved against.

The actions and policies of politicians and bureaucracies now reflect the persistent beliefs of sociologists that secularization, meaning the removal of religious influence overall, is in fact inevitable and, because inevitable, somehow a good thing. The pretence of neutrality both on the theoretical and now the practical level has been dropped. Secularism has now evolved a content or an ideology that can be seen in state support for population control, the formation of common attitudes regarding health issues including cigarette smoking and obesity, inclusion of other cultures and denigration of "Western civilization," defending transgressive types of sexual behavior, and encouraging obedience to bureaucratic authority. Since many or most college-educated people think, not unreasonably in most cases, that these are good ends for the state to pursue, national governments do pursue them.

But to what end? State secularism lies at the end of the journey of trying to discover and uncover the causes of the evident discordance of American culture. This *disunity* is troubling to us as Americans, since it lies as the cause of much personal dissatisfaction and even more of social conflict. On the other hand, given the increasing power of the bureaucratic state on all levels, perhaps, as this chapter will consider, such disunity is not necessarily a bad thing in the present case.

The Nature of the New Secular State

Having introduced three categories to describe secularization—the partial, complete, and state forms—and looking back on the facts and argument of this book, the final chapter should contain a description of the state of the culture as expressed in the arrival of the new secular state.

Secularism is a "mood" in society, one of those seemingly unresistable social movements or ideas, a kind of tsunami that overtakes societies and cultures from time to time; Burke said that to oppose them seemed not only futile but mischievous. Thus the new mood of secularism is not strictly the result of science overcoming religious faith, but something else in essence, since scientific ideals can no longer provide an intellectual justification for the modern state.

Nonetheless, secularism has now been incorporated in the politics of the new, secular state. In America lately, certain groups such as the American Civil Liberties Union have gone to courts in case after case in an attempt to remove every example of religious expression in any remotely public space, an effort that has been encouraged by the courts themselves. Neuhaus famously defined the phenomenon as the "naked public square," recounting the negative effects of explicitly secular ideals replacing traditional ideals based on a strong element of informal Christianity.

Yet there is a response by advocates of the new secular state, for they promote a notion of a public policy to which all citizens voluntarily adhere, by means of persuasion and the rational argument. This public policy, it is argued, is the basis for a society in which by implication, all citizens agree to participate, by means of a social contract in Rawls's version, since it provides the moral and intellectual basis for social participation. In this idea of public policy, citizens must, as it were, leave their most deeply held convictions at the door, for they are relegated as a matter of private interest only. It is all one, no matter what those convictions are, whether religious or rationalistic, libertarian or socialist, atheist or biblical; the only requirement is that such beliefs do not impact other individuals or the society in general—con men and murderers need not apply for citizenship under this rubric. Yet this notion is unrealistic for two reasons. First, the difference in the ideals expressed in a very diverse culture are antagonistic and sometime hostile to one another, which means that underneath the mask of social manners by which we engage in such transactions as shopping, selling, buying and working, and living in neighborhoods, there is often an outright anger that cannot always be disguised. The political and cultural polarization that the nation currently experiences is the result. Second, the state must rest on a positive view of itself that is held by the people at large, if individuals are to participate fully in it and to support it. It requires outright suppression then of expressions of personal belief and a kind of soft tyranny as the propaganda and legal apparatus of the secular state rain against its critics and opponents of its increasingly controversial policies.

With regard to morality, the newly emerging secular state is uninhibited by religious doctrine that promotes personal morality, especially those regarding sex, family life, and marriage. The new state's moral instructions to the individual citizen are to have a concern for the protection of individual rights, including their own and others, and to live responsibly according to a strictly utilitarian ideal; the chief secular ideals seem to be concern for the environment, healthful living (the secular state monitors psychological as well as physical health), equality of condition both economic and social, and toleration of differences. Existential questions about meaning and ultimate purpose in life are relegated to personal beliefs that are to have no effect on the body politic as the mere expression of such concerns amounts to unwarranted interference. Necessarily, religion in general and religious believers as a group are relegated to a narrowly restricted space so that they will have as little influence as possible.

The new, secular state is uninhibited by the findings of objective science, which can be interpreted to contradict its declared policies. ISCR has provided the chief example, but the scientifically obtainable evidence regarding the effect of children being raised by a single parent, for another example, demonstrates the general ability of the secular state to take what reinforces its policies and deny or ignore what contradicts them from current scientific research. Naturally, the political opponents of state policies will seize on the science that avatars of the state policies have ignored with the result that on particular issues (e.g., one-parent families, the rise of STDs, and global warming), *science becomes politicized*, as disagreements within the scientific community fall into the partisan divide characterized by the blue-state red-state dichotomy.

The new, secular state promotes non-reproductive sex by providing free access to abortion and birth control as well as by legalizing homosexual unions. This amounts to a deliberate policy as the state now intends to provide succor and support formerly given to persons by parents, families, and intermediate social entities. All express sentiment, sense of mutual obligation and social bonds that are the basis of a person's attachment to other people, families, local environments, and voluntary associations are thereby weakened.

The final observation is that the new secular state is in an intellectual or ideological condition of contradiction. The implied philosophy of the new state is utilitarian, but it relies on non-utilitarian ideals it will not promote. As a result, the new secular state is founded on the intensity of its propaganda, since neither faith nor reason supports it. Where there is no vision, it is said, the people perish; however, the new state deliberately denies even the possibility of a vision. The new secular state is not only coldly uninspiring, it is actively hostile to the search for ultimate meaning in life, because once brought into the public square, rationality and faith upset its anodyne ideal of public life. As can be seen from demographic considerations alone, the new secular state is unstable and will ultimately fail. The new state relies on no emotional or natural bonds except that of utilitarian necessity that leads to dependency but not devotion. It relies for its active support on a population it is encouraging to self-suppress, politicizing science while relying on technology, and relegating religious believers to second-class citizenship.

The Virtue of Cultural Disunity

Cultural conflict isn't so bad as long as there is a degree of freedom within the culture as a whole. In this way, the tectonic plates of oppositional opinions—scientific-based and faith-based ideals that have been examined here, for example—can move above and below one another when they come in contact, as they invariable will. Then the sense of disunity, expressed in various parts of the culture in disagreement with each other, can itself become a positive factor in the general sense of living in a democracy. We do not in this case freeze up into a defensive posture, but rather we await patiently, it is to be hoped, the fullest expression of an oppositional point of view. Then tolerance and freedom become touchstones and, as it is often expressed, respect for the opinion of others. But central to this democratic ideal is the denial to ourselves, on behalf of our scientifically or religiously based point of view, that we are ever likely to succeed in dominating the discussion by completely invalidating the opposition. Science, not only in its applications but in its ideals of first comprehending the physical universe by means of discoveries, laws, and rational theories, and then manipulating it for the good of mankind is not only not going to disappear but will remain forever essential to civilized life. Likewise, religious faith is not doomed to be overcome by scientific rationalism, no matter how loud and insistent faith's opponents become, even to the point of beginning to sound irrational in their attacks. But if, to return to the tectonic image, the plates in cultural opposition do not relent when they come in contact, and refuse to waver or slide smoothly around each other, a buildup of tension can occur, resulting in a cultural earthquake.

The religion-science conflict however, has been subsumed to a large extent, I have argued in Chapter 14, by the advance of technology and to a large degree by the failure of scientific explanation to explain ultimate questions. To be clear, the "failure" of science is not because physics, biology, and social science fail of their object of explaining physical, biological, or in every case, social phenomena. However, they do fail in providing a complete explanation of human life in the existential sense, of providing succor in personal tragedies or proving or disproving that there is life after death, and so on. Of course, scientific research and the empirical nature of scientific discovery are not intended to provide such broad explanations of philosophical and theological questions. However, it seemed necessary to make the point in some detail in the foregoing chapters, to offset the unexamined assertions of those who continue to loudly proclaim that science can explain it all for you.

But this "failure" of science in general to provide what it is never seriously intended to provide, namely an explanation of moral and religious issues, leaves a nettlesome problem in the context of the development of the new secular state and of its origins. For the origins of what we are referring to as the "new secular state" is not based on scientific explanations or on the results of scientific research, but on the broad idea that science is the exemplar of rational thinking. In this manner, the justification is so general and frankly vague that the scientific basis of the secular state cannot easily be disproven; the scientific basis for state policies is, as in the example of the debate surrounding ISCR, simply assumed or taken for granted. Further, science, as it does research that impacts social and political concerns is likely to become politicized, as it has recently in the global-warming controversy. In effect, the avatars of the secular state pick and choose which set of research results to use as a justification for policies that it has previously been set on. The relevant fact that stem cells obtained from the patient herself or from an umbilical cord have been much more effective than infant stem cells in providing therapeutic relief is simply ignored once, that is, the pursuit of infant stem cell research has become official state policy.

What then are the intellectual bases of the new secular state? It cannot be religion even in its most indirect manifestation, since such are feverishly rejected in every instance by the avatars of state secularism. But as I hope to have shown, it cannot be science either, since science is not understood in terms of its actual research results or the content of its theories to provide the intellectual and moral justification of the new secular state. Which leaves us with the perhaps obvious fact that the real justification for the new secular state is the acquisition of state power. The appearance of what is termed "the nanny state" is really insufficient to describe the effect of the increase of state power over individual lives; indeed it is too benign, as it implies that there is no more than a fussy but well-meaning interference. In reality, the mean-spirited attempt of the "nanny state" to control the minutia of our personal lives is an expression of Orwell's vision of the future of state power in *1984*, that of a boot planted for eternity in someone's face. It is, to return to the tectonic image, as if the state in its desire for absolute control has placed a network of titanium-steel lines around the plates, drilling down deep into the plates to anchor the steel chains, and thus give a sense of control over all the contending ideals and multitudinous personal behaviors that influence contemporary culture. But even a network of titanium steel cannot really control the movement of tectonic plates, which move on their own regardless of the attempt to control them. Ultimately, the effort is unworkable and as the plates continue their self-regarding and inevitable motions, the network will eventually break apart with probably explosive results.

The new secular state is unstable since it lacks a moral and intellectual basis; it is in perpetual war against religious believers, even though they constitute the most law abiding and faithful group of citizens in the state; it deliberately corrupts science to make it become the enabler of social control and the justification for unscientific and sometimes irrational policies, tyrannically pursued. In the end, the overall strategy of the new secular state is self-defeating because it discourages individual self control, encourages dependence, depresses the growth of the population among the better educated, and separates its cadre of decision makers from the sensibility of and social contact with the people it rules.

On Toleration, Religion, and Science

There is no shortage of intolerant attitudes swimming in the culture these days, and as a result, the virtue of toleration is highly touted; but state-supported ideals of tolerations seem to work only in one direction. Thus there is present a strong element of intolerance in the secular state's promotion of a necessary and essentially benign ideal of toleration, a phenomena is coming into notice in which the avatars of tolerance shut out serious criticism of the effects of state policies as they affect all segments of the general population. That is, under the encouragement of new state policies and outright propaganda from officially recognized groups, "toleration" implies that a passive majority should remain unresisting in the face of the new state's effort to reorganize society in a new and largely radical form (i.e., one in which the seemingly natural bonds that connect people to one another and individuals to a range of small and large social groups, from friendships and family, to region, city and state, is to be dissolved). The new secular state seems intent on replacing such attachments and reducing the relationship of individual to society to one type of relationship, that of the individual to the state—one size fits all. This point has been made repeatedly over the last two generations by a succession of critics and scholars.[1]

But to the point about toleration, toleration becomes in this context a means of suppressing debate and disarming resistance to new state policies, but in particular as it affects both religion and science. The conflicts between the policies promoting non-reproductive sex and traditional minded religious authorities are well known; the response of the new state is to enact policies that pose an outright challenge to religious authorities including entire denominations, and amounts to a threat to religious freedom for the individual believer. Bishops, preachers, and lay religious critics are now accusing the new secular state of systematic suppression of religious expression, that is, of not allowing religious expression beyond the privacy of one's own home or within the walls of the church, synagogue, or mosque. The issue is who will have influence in the public square; the doctrine of toleration is a means of undercutting the full expression of religious belief, particularly as it affects the secular state's promotion of non-reproductive sex.

Less visible but insidious is the inner conflict of the new state with scientific rationality. Because the state will pick and choose which scientific research results to promote and recognize and which to ignore—and defame, and also because the state holds the very large amounts of funding now required to do scientific research, the new state has a very powerful influence on the process of scientific research.[2] Politicians will not approach the tentativeness of scientific research results with the caution they deserve, but will instead reduce scientific research issues to partisan dichotomies, while any apparent ethical ramifications are ignored (religious objections having been put out of court). Technological application and the utilitarian results of the happiness of the majority and ease and comfort of the voting public now override the ideal of science as an exploration of the inner laws of the physical universe. The new state will surely be intolerant toward religious activities that interfere with the enacting of its policies, but intolerant as well toward scientific research whose results do not support its policies.

Fighting intolerance has long been encouraged but how to apply it to the intolerance of dissident ideas now found in America's high culture? The prime mover of such attitudes of intolerant tolerance are the academic institutions of the culture, and if we need to encourage a more open attitude it would be here in the academy where discussion, development, and dissemination of ideas usually takes place. That academic attitudes not merely reject but actively suppress discussion about reproductive sex, traditional religious beliefs, the market system, Western civilization, American exceptionalism, or the benefits of traditional family life is a demonstrable fact of long standing. Not only the criticisms of the academy by such ferocious critics as David Horowitz attest to the suppression of dissident voices in the academy. There stands the manifest fact that proponents of right-wing points of view have, over the last two generations, had to establish think tanks, independent institutions financed by private donations that support research and scholarship that can find no place in the university system. Such institutions as the American Enterprise Institute, the Heritage Foundation, Ethics and Public Policy Institute, the CATO Institute, and the Intercollegiate Studies Institute have put forth studies and reviews that helped to develop policies for Republican administrations. As importantly, they have provided an outlet for the work of dissident, independent scholars (including the present author, courtesy of ISI publications) and a safe harbor for full-time scholars not welcome in the academy. The list of important scholars working at think tanks includes George Weigle, Russell Kirk, and Michael Novak, among others who either were not thought to be creditable enough scholars to be approved for professorships or did not feel they would be welcome at America's colleges and universities. If any American university were to conclude that it should broaden its scholarship by including dissident right-wing thought, it could best begin by "raiding" the Washington right-wing think tanks.

Notes

1. See Robert Nisbet, *Twilight of Authority.*
2. Daniel S. Greenberg, *Science, Money and Politics: Political Triumph and Ethical Erosion*; Chicago, U. of Chicago Press, 2001.

Index